Lecture Notes in Electrical Engineering

Volume 1540

The book series *Lecture Notes in Electrical Engineering* (LNEE) publishes the latest developments in Electrical Engineering—quickly, informally and in high quality. While original research reported in proceedings and monographs has traditionally formed the core of LNEE, we also encourage authors to submit books devoted to supporting student education and professional training in the various fields and applications areas of electrical engineering. The series cover classical and emerging topics concerning:

- Communication Engineering, Information Theory and Networks
- Electronics Engineering and Microelectronics
- Signal, Image and Speech Processing
- Wireless and Mobile Communication
- Circuits and Systems
- Energy Systems, Power Electronics and Electrical Machines
- Electro-optical Engineering
- Instrumentation Engineering
- Avionics Engineering
- Control Systems
- Internet-of-Things and Cybersecurity
- Biomedical Devices, MEMS and NEMS

For general information about this book series, comments or suggestions, please contact leontina.dicecco@springer.com.

To submit a proposal or request further information, please contact the Publishing Editor in your country:

China

Jasmine Dou, Editor (jasmine.dou@springer.com)

India, Japan, Rest of Asia

Swati Meherishi, Editorial Director (Swati.Meherishi@springer.com)

Southeast Asia, Australia, New Zealand

Ramesh Nath Premnath, Editor (ramesh.premnath@springernature.com)

USA, Canada

Michael Luby, Senior Editor (michael.luby@springer.com)

All other Countries

Leontina Di Cecco, Senior Editor (leontina.dicecco@springer.com)

**** This series is indexed by EI Compendex and Scopus databases. ****

David Borge-Diez · Enrique Rosales-Asensio ·
Francisco José García-Moya

Fundamentals of Electrical Power Engineering

From Theory to Machines

Springer

David Borge-Diez
Electrical, Systems and Automation Engineering
University of Leon
León, Spain

Enrique Rosales-Asensio
Department of Electrical Engineering
University of Las Palmas de Gran Canaria
Las Palmas de Gran Canaria
Las Palmas, Spain

Francisco José García-Moya
Electrical, Systems and Automation Engineering
University of Leon
León, Spain

ISSN 1876-1100 ISSN 1876-1119 (electronic)
Lecture Notes in Electrical Engineering
ISBN 978-3-032-18903-5 ISBN 978-3-032-18904-2 (eBook)
https://doi.org/10.1007/978-3-032-18904-2

This Springer imprint is published by the registered company Springer Nature Switzerland AG
The registered company address is: Gewerbestrasse 11, 6330 Cham, Switzerland

Preface

The purpose of this book is to review concepts that students will face in a first course in an electrical engineering curriculum. This book is divided into 10 chapters. Chapter 1 provides a comprehensive review of electricity, and electrical engineering. This chapter differentiates between the applications of electricity, which we call electrical engineering and electricity as a fundamental force of nature. This chapter also focuses on the electrokinetics, as well on the common elements of an electric circuit. Finally, the authors provide an outlook on the influence of temperature on the electric current.

Chapter 2 provides an overview of the main law for electric circuits problem-solving, Ohm's law. The authors focus on the analysis of the different ways for grouping resistances. It also provides a definition of the DC generator including the final voltage as well as the internal resistance. Thus, the study of the generator behavior is of great importance for electricity applications. Also, the authors provide an outlook on the basis of coupling between generators, and between electric receivers. Finally, the LTspice tool for solving circuits is presented by means of several examples to simulate electric circuits.

Chapter 3 describes the main features of passive components in an electric circuit including resistors, capacitors, and inductances. In this chapter, we study the behavior of these components as well their characteristics. The grouping conditions were described. Finally, this chapter also provides an outlook on the RL and RC circuits.

Chapter 4 presents the Ohm's law in sinusoidal steady state through the analysis of the purely resistive circuit, or in a purely capacitor circuit. After the analysis of a purely inductive circuit, it presents Faraday's law. This chapter finishes analyzing the sinusoidal RL, RC, and RLC series circuit.

Chapter 5 focuses in the definition of power. Especially it focuses in analyzing the power in a purely resistive circuit, in circuits with pure coil, and in a purely capacitive circuit. It also reviews power in a RL series circuit.

Chapter 6 classifies electrical machines. It presents the constituent elements of rotating machines and reviews the process of conversion of energy in electrical machines. It analyzes the magnetic quantities, units, and basic laws of electromagnetism in ferromagnetic materials. By analyzing the Barlow wheel, it studies the application of the fundamental laws. The electrical circuits of electrical machines are reviewed analyzing the winding factor, distribution factor, pitch factor, and slot tilt factor. It finishes presenting the basic generator and the values associated with an alternating sine wave.

This Chap. 7 analyzes power transformers. After introducing their function and working principle, it reviews the parts, and losses of power transformers. This chapter also defines the single-phase power transformer, voltage regulation, or transformer efficiency. Finally, this chapter reviews different operating conditions of the transformer and the types of three-phase transformers.

Chapter 8 introduces the power of the three-phase system including star balanced load and balanced load in delta. In addition, the efficiency of the electric motor is defined in this chapter.

Chapter 9 starts reviewing electrical power system concepts. After that, it proposes a classification of asynchronous machines. It continues reviewing the constituent parts and their function. In addition, the operating modes are analyzed. Finally, this chapter provides an outlook of the electromagnetic theory including the Ferraris theorem, the characteristic curves, and torque study of asynchronous machines.

Chapter 10 provides an overview of the synchronous machine. It reviews the constituent elements including the rotor, stator, as well the cooling system, or the outer casing. In the next step it presents the different types of synchronous machine and analyzes the operating conditions of synchronous generators. This chapter finishes analyzing the power factor correction.

Las Palmas de Gran Canaria, Spain — Enrique Rosales-Asensio
León, Spain — David Borge-Diez
León, Spain — Francisco José García-Moya

Competing Interests The authors have no competing interests to declare that are relevant to the content of this manuscript.

Contents

Chapter 1
Electrical Engineering: Contextualization and Applicability

Abstract Electricity is the science that studies the natural phenomena we call electrical. From the study of this phenomenon have arisen many applications, which include: electric light, electric machines, motors, generators, computers, etc. The study of the many applications of electricity laws is what we call electrical engineering. In an electrical circuit there are active and passive elements. This chapter also analyzes the characteristics of materials regarding electrical conduction.

Keywords Electricity and electrical engineering · Electrostatic and electrodynamics · Magnetic forces · Electricity applications · Conductors and isolators

1.1 Introduction

In this chapter, we review the model that allows us to study the problem of electrical machines with elements that make up electrical circuits and the associated quantities. It is common that electric or electronic engineering students separately start learning electrostatics, electrokinetics, magnetostatics, and electro magnetism in order to gain a complete understanding of the phenomena regarding to electricity and magnetism. For a deep understanding of these concepts please refer to a complete electromagnetism foundations course.

Warning Electricity VS Electrical Engineering.

It is challenging to distinguish clearly between the content of electricity and that which pertains to electrical engineering. This is because understanding electrical phenomena is necessary to apply their governing laws. The field of electricity entails knowledge of these phenomena to comprehend the world around us. Electricity is modeled with equations that define the phenomena. Specially, it highlights the Maxwell equations for modeling the electromagnetic phenomena. Conversely, electrical engineering focuses on applying these equations for practical applications. The field of electricity encompasses the study of the phenomena resulting from electric

D. Borge-Diez et al., *Fundamentals of Electrical Power Engineering*, Lecture Notes in Electrical Engineering 1540, https://doi.org/10.1007/978-3-032-18904-2_1

charges and their interaction. Nevertheless, in order to understand the electrical engineering rely on, this book reviews the most outstanding concepts and equations that describe interactions that occur in the widespread electrical machines that we daily interact with.

Definition (*Electrostatics VS electrokinetics*) Electrostatics is the branch of electrical engineering that investigates charged particles at rest. It also encompasses the study of electric fields that do not change over time. In contrast, electrokinetics is the branch of physics that deals with the motion of charged particles.

Definition (*Electric force*) It is one of the "fundamental forces of nature" that exists between bodies bearing electric charges that are not equal. Bodies with unlike charges experience mutual attraction; those with like charges repel mutually. Electric charges and magnetic monopoles provoke fields, **E**, **B** (Fig. 1.1) either electrostatic and magnetostatic or electromagnetic.

Definition (*Coulomb's Law*) Coulomb's law states that the attracting or repulsive electrostatic force between two charges is directly proportional to the product of their magnitudes and inversely proportional to the square of the distance between them (Eq. 1.1). When a charge is stationary, or static, it produces electric forces on other charges in its region of space.

$$\mathbf{F} = k\frac{q_1 \cdot q_2}{r^2} \tag{1.1}$$

Being the proportionality constant, k, often called Coulomb's constant, depending on the medium. The vector form of the electrostatic field between charges of the same polarity is repulsive (Fig. 1.2).

$$k = 9 \cdot 10^9 \mathrm{N} \cdot \mathrm{m}^2 \cdot \mathrm{C}^{-2}$$

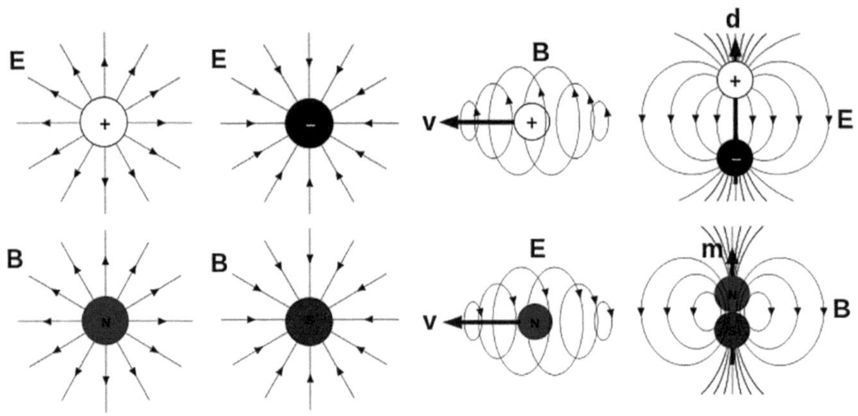

Fig. 1.1 Fields caused by electric charges and magnetic monopoles. From wikicommons under free to use license [1]

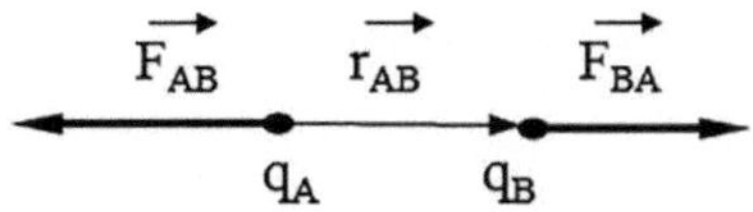

Fig. 1.2 Repulsive electrostatic force between two charges. *Source* Own elaboration

1.1.1 Force Exerted in a Magnetic Field

The force acting on a charge q that does not move when locates inside an electric field then appears a force (Eq. 1.2).

$$\mathbf{Fe} = q \cdot \mathbf{E}(N) \tag{1.2}$$

The electric force depends on the position of the charge q, the electric field that affects it and the magnitude of the charge q as well. We also determine the magnetic field by the force that affects to a charge in motion.

In addition, the units we use to measure magnetic flux density, B, are: Wb m $^{-2}$ or T; 1 Gauss = 10 $^{-4}$ T.

$$[B] = \frac{[\mathbf{F}]}{[\mathbf{v}][q]} \rightarrow 1T = \frac{N}{C \cdot m \cdot s^{-1}} = \frac{N}{A \cdot m}$$

A charge in motion experiences other force, **Fm**, which is proportional to the charge, and depends on the velocity that the charge moves and the magnetic flux field as seen in (Eq. 1.3).

$$\mathbf{Fm} = q \cdot \mathbf{v} \times \mathbf{B}(N) \tag{1.3}$$

Then, (Eq. 1.4) illustrates vector force that the charge q experiences within the electric and magnetic field.

$$\mathbf{F} = q \cdot \mathbf{E} + q \cdot \mathbf{v} \times \mathbf{B} = q \cdot (\mathbf{E} + \mathbf{v} \times \mathbf{B})(N) \tag{1.4}$$

We see that a charge in motion also produces magnetic effects (Fig. 1.3). The direction of the magnetic field B, as well as the direction and speed of the charge q define the magnitude and direction resulting force.

The force direction indicates a negative charge moving clockwise, or a positive charge moving counterclockwise, consistent with the right-hand rule. The force this charge experiences is the Lorentz force that changes, in both magnitude and direction under different speed directions (Fig. 1.4).

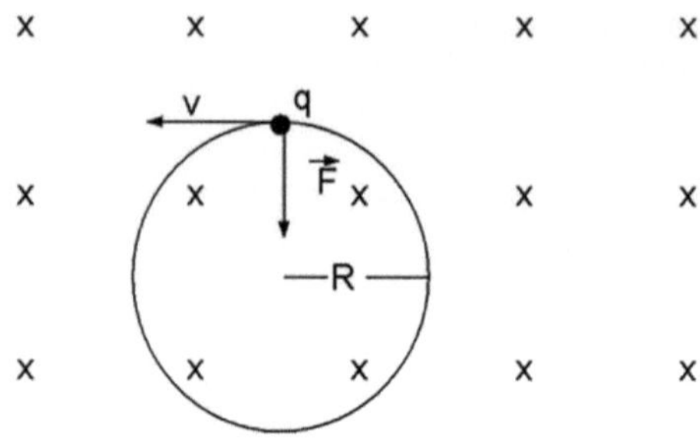

Fig. 1.3 Force exerted on a charge in motion. *Source* From wikicommons under free to use license [2]

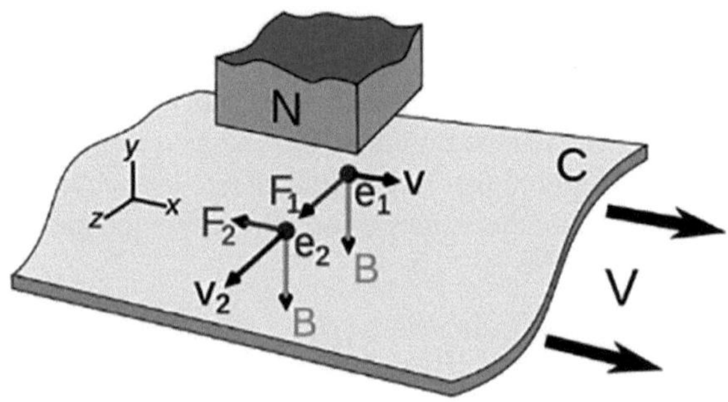

Fig. 1.4 Lorentz's force. *Source* From wikicommons under free to use license [3]

1.2 The Foundation of Electromagnetism: Maxwell Equations

Maxwell's equations are a set of four fundamental laws that completely describe the behavior of electric and magnetic fields. Formulated by James Clerk Maxwell in the nineteenth century these equations unified electricity, magnetism, and light into a single theoretical framework, laying the groundwork for modern physics and engineering. They are essential in understanding electromagnetic waves, circuits, and even modern technologies like wireless communication and optics.

1.2.1 The Four Maxwell's Equations

Maxwell's equations can be expressed in both differential and integral forms, each providing insights into the nature of electromagnetic fields.

1.2.1.1 Gauss's Law for Electricity

Gauss's law states that the electric flux through a closed surface is proportional to the total charge enclosed within that surface. It describes how electric fields originate from electric charges.

Equation (Differential Form):

$$\nabla \cdot \boldsymbol{D} = \rho_V$$

Equation (Integral Form):

$$\oint \mathbf{D} \cdot d\mathbf{s} = Q$$

1.2.1.2 There is no Magnetic Isolated Charge

Equation (Differential Form):

$$\nabla \cdot \mathbf{B} = 0$$

Equation (Integral Form):

$$\oint \mathbf{B} \cdot d\mathbf{s} = 0$$

The importance of this law relates to the meaning of this equation: there is no isolated magnetic charge. This law asserts that there are no magnetic monopoles—magnetic field lines are continuous and always form closed loops. Unlike electric fields, magnetic fields do not diverge from a single point.

1.2.1.3 Faraday's Law of Induction

Equation (Differential Form):

$$\nabla \times \boldsymbol{E} = -\frac{\boldsymbol{dB}}{\boldsymbol{dt}}$$

Equation (Integral Form):

$$\oint \mathbf{E} \cdot dl = -\frac{d\phi}{dt}$$

Faraday's law explains how a changing magnetic field induces an electric field. This principle is the basis for electric generators and transformers.

1.2.1.4 Ampère's Law with Maxwell's Correction

Ampère's original law related magnetic fields to electric currents. Maxwell added the displacement current term, which accounts for time-varying electric fields, completing the symmetry with Faraday's law.

Equation (Differential Form):

$$\nabla \times \boldsymbol{H} = \boldsymbol{J} + \frac{d\mathbf{D}}{dt}$$

Equation (Integral Form):

$$\oint \mathbf{H} \cdot dl = \mathbf{I} + \int \frac{d\mathbf{D}}{dt} \cdot d\mathbf{s}$$

Additionally, Maxwell's equations predict the existence of electromagnetic waves, which travel at the speed of light. This led to the realization that light itself is an electromagnetic wave, unifying optics with electromagnetism. Maxwell's equations describe Radio, Wi-Fi, and mobile networks rely on electromagnetic wave propagation. From them we derive equations that let us design circuits, antennas, and power systems. It also allows us to describe the behavior of light in lenses, lasers, and fiber optics. Finally, they inspired Einstein's theory of special relativity, as Maxwell's equations are invariant under Lorentz transformations.

Maxwell's equations are one of the greatest achievements in physics, providing a complete description of classical electromagnetism. They not only explain electrostatic, but also electromagnetism as one body itself, forming the basis for much of modern technology. From power generation to telecommunications, their impact is immeasurable, demonstrating the profound connection between mathematics and the physical world.

For what electrical, or electronic engineering concerns, if we focus in reviewing these equations when applying them to the varied shapes, materials that we normally find in these devices, it would be almost impossible to figure out a solution. Throughout those years, it has been developed the complete theory of electrical engineering to simplify the calculus of what is going on in those electromagnetic devices. This is done by means of creating models that truly represent what the complete theory specifies. The lumped circuit abstraction highlights as a simplified representation of a physical system or circuit that assumes components such as resistors, inductances, among others are concentrated at a single point. Then, engineers apply laws and theorems to solve these circuits. Not only Kirchhoff's, and Ohm's Laws, but also Thevenin and Norton's theorems, as well others are extensively used to analyze electromagnetic devices. From energy generation to consumption in devices, this abstraction allow us to model the behavior of electromagnetic devices.

The main focus of this book is to review the basic concepts of electrostatics and electrodynamics with the aim of understanding how main electric machines work. With these premises, we analyze the nature of electricity and how an electric current can be generated by means of moving devices.

1.3 Kirchhoff's Laws and Their Connection to Maxwell's Equations

Kirchhoff's laws—the Current Law (KCL) and the Voltage Law (KVL)—are fundamental principles in circuit theory, governing the behavior of electric currents and voltages in networks. Kirchhoff's laws are simplified versions of Maxwell's equations, valid under low-frequency and lumped-element conditions. While these laws are widely used in electrical engineering, they are actually special cases of Maxwell's equations under certain assumptions.

1.3.1 Kirchhoff's Current Law (KCL)

The sum of currents entering a junction in a circuit equals the sum of currents leaving it. We express it in its mathematical form as:

$$I_{in} = \sum I_{out}$$

The physical meaning of this equation is due to KCL is a consequence of the conservation of electric charge. Since charge cannot accumulate at a point in a steady-state circuit, the net current flow into any node must be zero.

1.3.2 Kirchhoff's Voltage Law (KVL)

Kirchhoff's Voltage Law states that the sum of voltage drops around any closed loop in a circuit is zero. Its mathematical form:

$$\mathrm{V_k} = 0$$

The physical meaning of KVL arises from the conservation of energy in an electric circuit. It ensures that the work done by electric fields around a closed loop is zero, meaning no net energy is gained or lost in a complete cycle.

Deriving Kirchhoff's Laws from Maxwell's Equations.

As we have previously mentioned Kirchhoff's laws are simplifications of Maxwell's equations under low-frequency and lumped-element approximations. We derive KCL from Gauss's Law and Charge Continuity. Taking into account that

$$\nabla \cdot \mathbf{D} = \rho_V$$

where **D** is the electric displacement field and $\boldsymbol{\rho}_{\mathbf{v}}$ is the volumetric density of free charge.

And charge continuity equation:

$$\nabla \cdot \mathbf{J} + \frac{d\rho_V}{d_t} = 0$$

where **J** is the current density and $\boldsymbol{\rho}_{\mathbf{v}}$ is the volumetric density of free charge.

In order to derive the KCL we assume that there is no charge accumulation (steady-state, $\partial\rho/\partial t \approx 0$). We also build the lumped circuit model where dimensions are much lower than wavelength, and then there is no significant electromagnetic radiation.

Under these conditions we can write that

$\nabla \cdot \mathbf{J} = 0$, and then $\mathrm{I} = 0$.

This is KCL.

KVL from Faraday's Law.

According to Maxwell-Faraday's law we can write that

$$\nabla \times \mathrm{E} = -\frac{\mathrm{dB}}{\mathrm{d_t}}$$

In order to build the model for analyzing circuits we assume for KVL that in static magnetic fields ($\partial \mathrm{B}/\partial \mathrm{t} \approx 0$) there is no induced EMF. In addition, we assume ideal conductors, then no resistive voltage drop outside components.

Then, we have:

$$\nabla \times E = 0$$

$$E \cdot dl = 0$$

This integral form means the sum of voltage drops around a loop is zero (KVL).

While Kirchhoff's laws are powerful, they fail in high-frequency circuits where $\partial \mathrm{B}/\partial \mathrm{t} \neq 0$, leading to induced EMFs. Then, we should not use them to analyze distributed systems such as transmission lines, or antennas. In addition, we use it in circuits with non-lumped elements, where wave propagation matters. In such cases, full Maxwell's equations must be used instead.

1.4 Constitution of Copper

Electric current commonly flows through wires. Copper (Cu) is one of the most widely used materials in the electricity industry. For this reason, we study its composition. Minerals such as azurite, malachite and others contain copper in different proportions. Different mineral contain cooper (Fig. 1.5), and we use different processes to extract cooper with various purity grades.

The atom of copper contains 29 protons and 35 neutrons. There are 29 electrons in different orbits around the nucleus (Fig. 1.6). The atom is neutral from the outside because each proton has a positive charge and each electron has a negative charge, both of which are equal. Since matter is made up of atoms, it is also neutral.

Fig. 1.5 Copper mineral. *Source* From wikicommons under free to use license [4]

Fig. 1.6 Electron shell of copper. *Source* Own elaboration

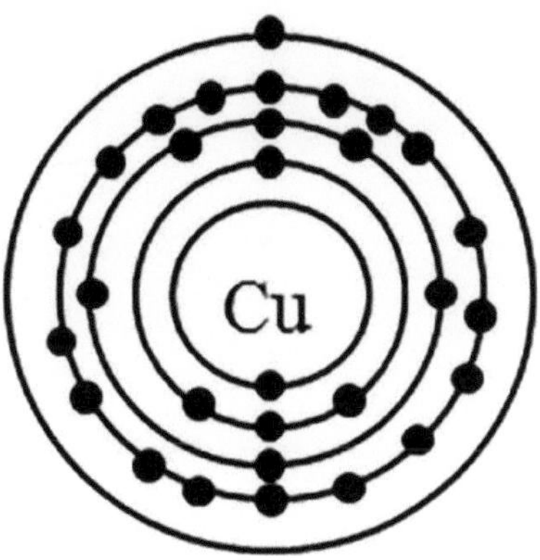

Two electrons traveling in circular orbits occupy the first energy level. These electrons are closely bound to the nucleus. This is due to the strong electrical attraction forces between protons and electrons at this energy level, which is very close to the nucleus. The second energy level contains 8 more electrons. These electrons follow circular and elliptical orbits (Fig. 1.6). The attractive forces between the protons in the nucleus and the electrons are smaller in this orbit. This is due to the greater distance between them. The energy of the electrons in this level is therefore higher than in the previous one. The third energy level contains 18 electrons that follow circular and elliptical orbits. Then, the fourth orbit is occupied by a single electron that is very weakly bound to the proton of the nucleus. This is what gives the copper atom its special electrical conductivity quality.

Once a shell or layer is complete, the electrons are tightly bound to the nucleus and it is very difficult to release them from their orbitals. In the case of copper, on the other hand, the last layer is not complete. In this case, electrons, called valence electrons, experience high mobility and they also can change their orbits without much difficulty. This is why we call them free electrons. In case of materials or elements containing these atoms, the electrons of the last layer (free electrons) are in constant motion, moving from one atom to another adjacent atom. Under an electromotive force, atoms constantly lose electrons and become a positive ion (Copper wire). The positive ions attract electrons and then an electron flow is established (Fig. 1.7). We have to decide the flow direction. This figure also shows the conventional current flow direction as the contrary of the flow of electrons.

When an atom loses an electron, it becomes positively charged. Since it has more positive charges, it becomes a positive ion. As a result, its state is no longer neutral because there is an imbalance of charges.

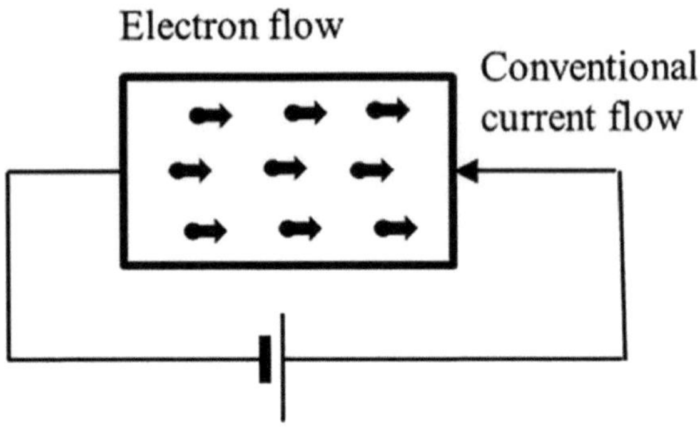

Fig. 1.7 Electrons flowing in copper. *Source* Own elaboration

1.5 Conductive and Insulating Materials

Depending on the structure of their atoms, materials can be conductors or insulators. All phenomena related to electrical conduction take place at the outer edge of the crust and between atoms. Conductors are substances whose final layer contains one or two electrons in an incomplete orbit. The atomic structure of good conductors usually contains only one electron in the outermost shell of the atom. This is called the valence electron, and it is easily released from the atom, allowing current to flow.

Question Then, when a material is more conductive?

We say that a piece of material is more conductive the freer electrons it has and the less they are bound to the nucleus (far orbitals). The most conductive materials are metals. Among them, the best conductors are silver, copper, gold and aluminum, in that order. Conductive materials allow electrons to easily flow through them. In this case, if we apply an electric force to them, a flow of electrons is established and we call it electric current. Different elements have different characteristics of electrical conductivity as well electrical resistance (Table 1.1).

On the other hand, we call insulators to materials that are made up of atoms with complete outer shells or missing one or two electrons. In these materials, the atoms hold on tightly to their electrons, which have little mobility. This makes it difficult for the electron to move through the atoms.

Question Which materials are good insulators?

Among the insulators, we highlight: glass, porcelain, and mica, silk, Sulphur, dry air, etc., which do not have such free electrons.

Question What happens when we apply very large electrical forces to insulators?

Table 1.1 Electrical conductivity and resistivity

Conductor	Electrical conductivity (10^6 Siemens/m)	Electrical resistivity (10^{-8} Ohm/m)
Silver	62.1	1.64
Copper	58.5	1.72
Gold	44.2	2.3
Aluminum	36.9	2.59
Molybdenum	18.7	5.34
Zinc	16.6	6
Tungsten	8.9	11.2
Brass	15.9	6.3
Lead	4.7	21.3
Iron	10.1	9.8
Mercury	1.1	95.8
Carbon	5.9	16.9

Source [5, 6]

When very large electrical forces are applied, an electric current can be obtained through the piece of material. This is because its atomic structure is broken, and it is the reason why there are no perfect insulators.

1.6 Electricity Generation

Electricity exists in all matter. However, matter is electrically neutral. Therefore, to establish an electric current some external energy must be applied in order to release some electrons (electrical charge), i.e. to obtain electricity. It can be done by means of using the mechanical energy in turbines, either wind, water, or others attached to synchronous, asynchronous, or other electric generators. We can model every device with the lumped circuit abstraction like the model of the solar cell that harvests the energy contained in sunbeams.

1.6.1 Electrodynamics

1.6.1.1 Elements of an Electric Circuit

First, if we want to establish an electric current in a circuit, we need a path for the electrons to move (electrodynamics). Then, an electric circuit must be closed in a model of a closed circuit containing at least a voltage source and a resistor (Fig. 1.8). An electrical circuit is the path that electrons take when there is no obstacle to their movement, i.e. when it is closed. The voltage source models a system that maintains an electric force. For electricity to flow, it must be in a closed circuit. The resistor models the behavior of the electrons in the wire. Electric current can only flow in a closed circuit.

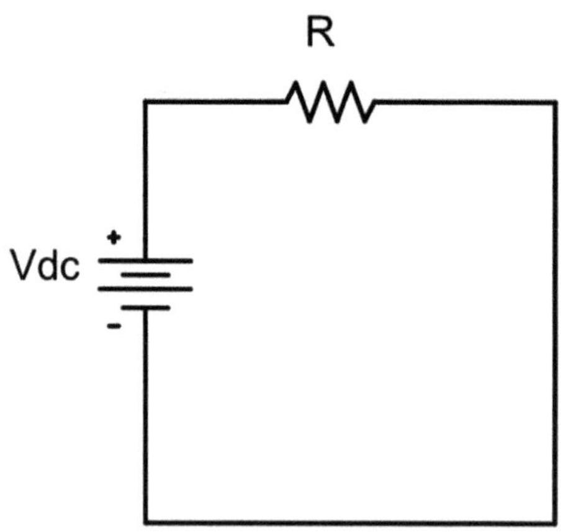

Fig. 1.8 Flow of electrons in a closed circuit. *Source* Own elaboration

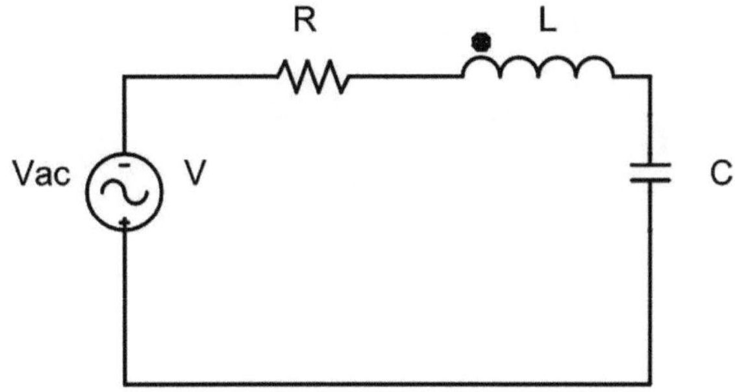

Fig. 1.9 Passive elements in a circuit. *Source* Own elaboration

An electrical circuit is the set of elements necessary to create an electrical current, as well as others capable of transforming this electrical energy into another type of energy. In a circuit we find active elements such as voltage or current sources. These elements can supply energy in different conditions. Among others, we can also find Resistors (R); Inductors (L) and Capacitors (C) (Fig. 1.9). These are known as passive elements. We will focus on the behavior of these passive elements.

The source delivers energy, and the passive elements convert it into another form of energy or to store it in the form of a magnetic or electric field. We build models with circuit elements (Fig. 1.10) according to the function they perform. Elements (a) and (b) are current sources. Elements (c) and (d) are voltage sources. Elements (e) are the electronic symbols for a resistor. The top is the American version. The bottom is the International Electro technical Commission version. Elements (f), (g), (h), (i) are the symbols for inductors. In addition, (j), (k), (l) represent capacitors.

When a voltage or current source is not affected by changes in the circuit which it is connected to, it is said to be an independent source. In this case it is represented by a circle. A dependent voltage or current source is one whose characteristics change according to the conditions of the circuit. In this case it is represented by a rhombus. The three passive elements, resistors, coils and capacitors are shown in figures (e), (f) and (g) respectively. We also model different voltage sources that can be found in electrical circuits (Fig. 1.11).

Throughout this chapter we will study circuits that use direct current generators, such as batteries, accumulators and dynamos, and alternating current generators, such as synchronous generators. In addition, as mentioned earlier, an electrical circuit contains a variety of drive, control and protection devices that allow us to control electrical current with maximum efficiency and safety for people, animals and equipment.

Then, we can say that in each electrical circuit we observe the following elements:

1. Generator: It is the one that generates or produces the electricity flow.
2. Wires and connectors: These are the materials that allow the passage of electrons. The most commonly used is the copper wire.
3. Control elements: They allow us to control the flow of electricity in the electrical circuits. The best known and used control element is the switch, which controls the flow of electricity or not.

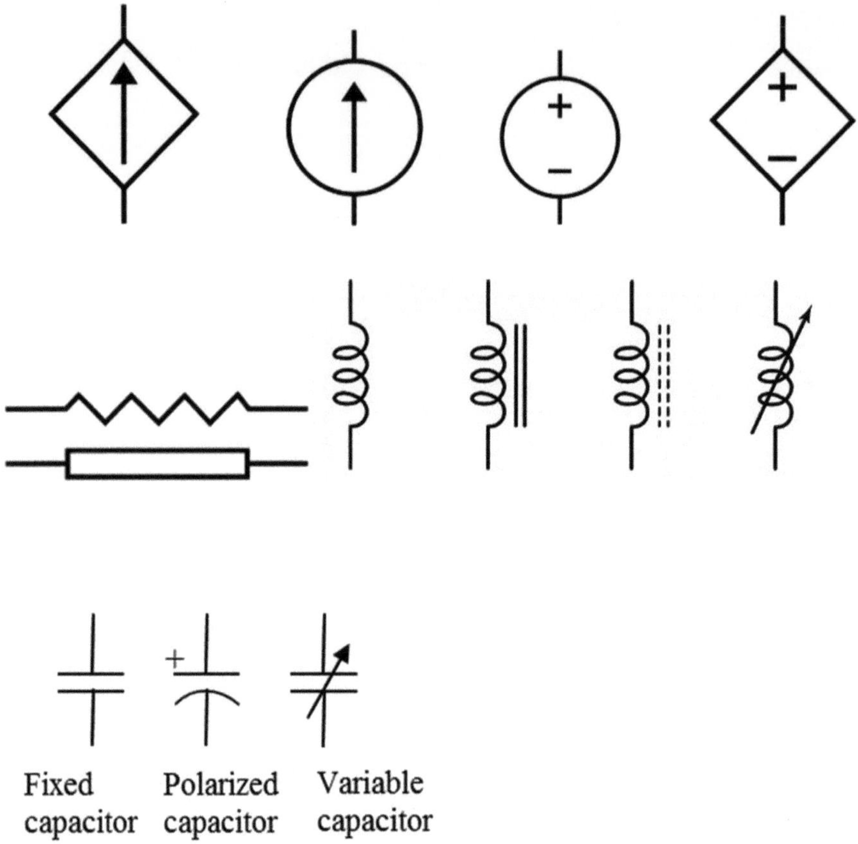

Fig. 1.10 Symbols for normalized elements. *Source* From wikicommons under free to use license [7]

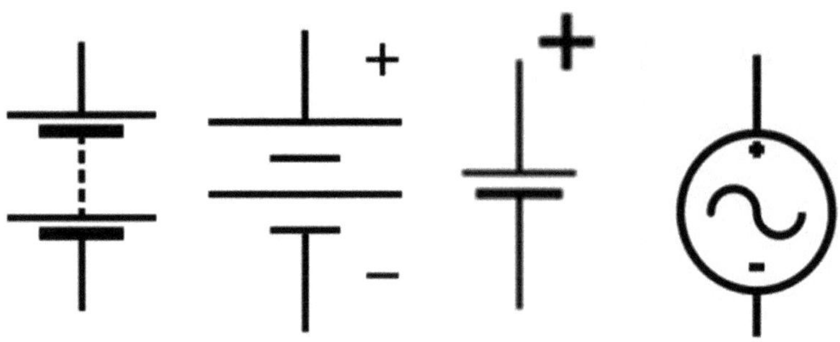

Fig. 1.11 Generator symbols. *Source* From wikicommons under free to use license [7]

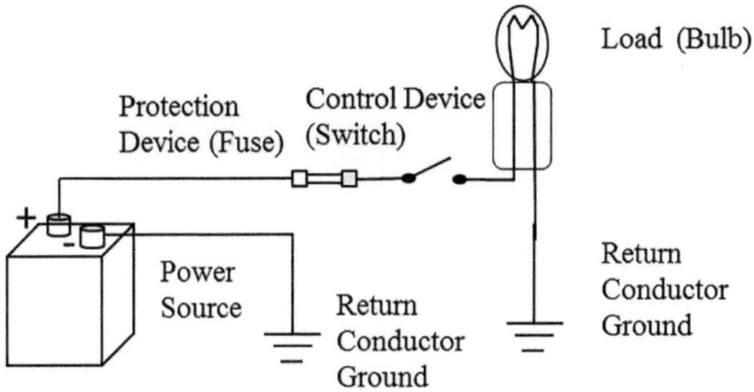

Fig. 1.12 Closed circuits in a vehicle. *Source* Own elaboration

4. Protections of the circuit elements: They are those that protect the circuit from overloads or short circuits. They also protect the user. For example, a fuse is an element that protects the circuit while the sheathing of copper wires is an element that protects the user.
5. Receiver or load: This is where the electrical energy is converted into another type of energy.

We see some of these elements, for example on self-propelled vehicles, where the negative terminal of the battery is connected to the chassis and engine block (Fig. 1.12).

This is done to close the circuit through the metal part of the car. The metal of the car is what we call the ground. This eliminates the return conductor.

1.6.2 Electric Current

The generator is the element of the circuit that provokes the electrons to flow into the circuit.

1.6.2.1 Current Density

We define the current density, vector **J** as the quotient of the current, I, and the area, S, of the conductor through which the current flows. Imagine the current as a bunch of electrons flowing: it the current density inside a conductor element, mainly wires (Fig. 1.13).

On the other hand, the rate of change of charge with respect to time through a surface, or current density is what we call current. It is common that the current flows in a circuit within a cylindrical wire's cross section (Fig. 1.14).

We can express both as

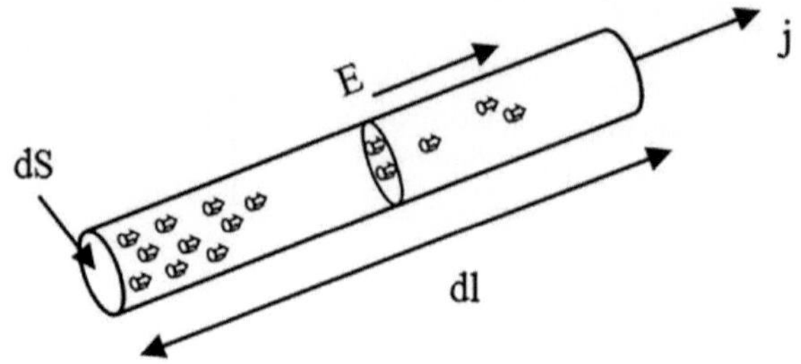

Fig. 1.13 Current density in a conductor. *Source* Own elaboration

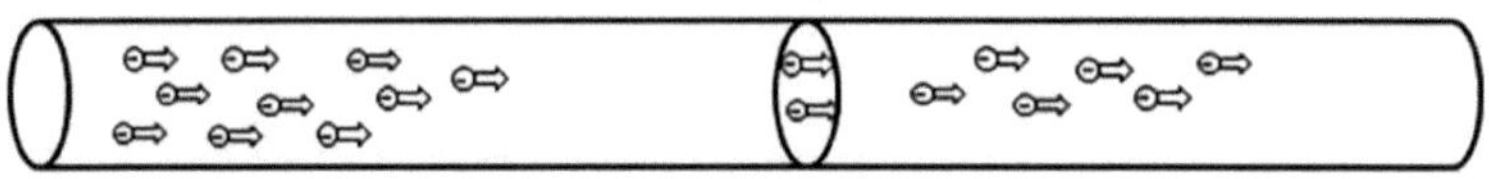

Fig. 1.14 Current through a cylindrical wire. *Source* Own elaboration

$$\mathbf{J} = \frac{I}{S}\left(\mathrm{A} \cdot \mathrm{mm}^{-2}\right) \tag{1.5}$$

where **J** is the current density in $\mathrm{A{\cdot}mm^{-2}}$, I is the current in A, and S is the area of the conductor through which the current flows (in mm^2)

It is important to highlight that the ability of materials to properly sustain the current flow differs. It is common to display this ability with the short-circuit current density G for materials like copper and other employed alloys (Table 1.2).

Many tables express maximum permissible currents in wires in $\mathrm{A/mm^2}$. We commonly take maximum permissible current from commercial catalogues and electrical standards for wires.

Table 1.2 Short-circuit current density Jsc (Max. Temperature 300 °C) [5, 6]

Time (s)	St/tZn ($\mathrm{A/mm^2}$)	Copper ($\mathrm{A/mm^2}$)	St/St (V4A) ($\mathrm{A/mm^2}$)
0.3	129	355	70
0.5	100	275	55
1	70	195	37
3	41	112	21
5	31	87	17

1.7 Electrical Resistance

1.7.1 Conductor Resistance

The resistance that a conductor offers to the flow of current depends on the length, area, and the type of material it is made of. We know that the path the electrons follow in a resistive medium is not linear (Fig. 1.15). Then, the resistance of a conductor varies not only according to the material but also to the dimensions of the material.

When modeling what happens in circuits, we have to represent the resistance that an element offers to the flow of current (Fig. 1.16).

The resistance of a piece of material is proportional to its length, resistivity, and inversely proportional to the cross-sectional area (Eq. 1.6).

$$R = \rho \frac{L}{S} \tag{1.6}$$

1.7.1.1 Resistivity

Resistivity is the opposition to the passage of electric current that offers a conductor of unit length and area. It is represented with the Greek letter rho, ρ.

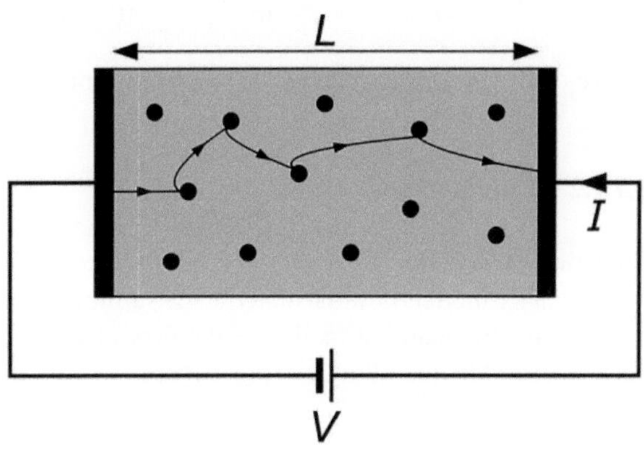

Fig. 1.15 Electrons scattering off atoms in a resistive medium. *Source* From wikicommons under free to use license [8]

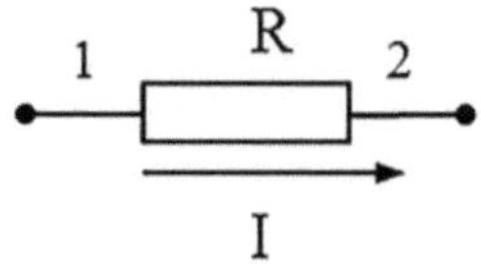

Fig. 1.16 Symbol for resistor. *Source* Own elaboration

Table 1.3 Copper typical values

Typical values		
Tensile strength	N/mm^2	220–300
Yield strength at 1% elongation	N/mm^2	120–200
Conductivity	$S\ m/mm^2$	58.5
Resistivity	$\Omega\text{-}mm^2/m$	0.0171
Thermal coefficient of resistance	$10^{-6}/K$	3900–4300
Density	Kg/dm^3	8.9

Source [9]

$$\rho = \frac{R \cdot S}{L} = \frac{\Omega \cdot \mathrm{mm}^2}{\mathrm{m}} = 10^{-6}\Omega \cdot \mathrm{m} \tag{1.7}$$

where ρ is the resistivity coefficient (Ω-mm^2/m), R is the electrical resistance in ohms (Ω), S is the conductor area in mm^2, and L is the conductor length in meters (m).

If we use the I.S. units, we have the ohm-meter as the unit of conductor resistivity in the I.S. According to its properties, the most used material for wires is copper (Table 1.3).

As we see, the resistivity coefficient of copper is 0.0171 (Ω· mm^2/m). This means that the resistivity of each meter is $1.71 \cdot 10^{-8}$ Ω m for a square millimeter of copper wire.

$$\rho = \frac{R \cdot S}{L} = \frac{0.0171\,\Omega \cdot \mathrm{mm}^2}{\mathrm{m}} \cdot \frac{1\,\mathrm{m}^2}{10^6\,\mathrm{mm}^2} = 1.71 \cdot 10^{-8}\Omega \cdot \mathrm{m}$$

We often use a resistivity of 1.810^{-8} Ω m for calculus with copper wires.

1.7.1.2 Conductance

Sometimes it is better to use reciprocal resistance, namely the conductance G. While the resistance models how the material resists the flow of electric current, conductance models how the material facilitates the flow of electric current through conductors. As a measurable quantity we express it as

$$G = \frac{1}{R}(S) \tag{1.8}$$

We often represent the conductance with the letter G, being the Siemens (s) the units of conductance. It is also expressed with the Greek letter mho (ohm spelt backwards), which is also represented rotating the symbol of resistance (Fig. 1.17).

For example, if resistance R = 10 Ω then conductance G = 0.1 S. Then,

$$G = \frac{1}{R};\ U = R \cdot I;\ GU = I$$

Fig. 1.17 Mho symbol. *Source* Own elaboration

Accordingly, we define the Siemens (S) as the conductance of a conductor having an electrical resistance of one ohm.

1.7.1.3 Conductivity

Conductivity is a macroscopic constitutive parameter of the material. Mathematically is the inverse of resistivity. Gamma, or sigma, is the Greek letter that stands for it (Eq. 1.9).

$$\gamma = \sigma = \frac{1}{\rho} = \frac{L}{R \cdot S} = \frac{\text{m}}{\Omega \cdot \text{mm}^2} = \frac{1}{\Omega \cdot \text{m}} = 1\,\text{Sm}^{-1} \tag{1.9}$$

where $\gamma = \sigma$ is the conductivity, in m/ (Ω·mm^2), ρ is the resistivity, in (Ω·mm^2)/m, L is the length, in m, R is the resistance, in Ω, and S is the area, in mm^2. We define conductivity as the ability of materials to conduct an electric current per unit length and unit cross section.

1.7.2 Influence of Temperature

The resistance of a conductor changes with temperature. If the temperature of a resistor changes, then so does its resistance. Normally, if the temperature of a resistor increases, then so does its resistance. The opposite happens when a conductor cools. The resistivity of copper and other metals is zero at absolute zero (–273 °C). The values previously presented (Table 1.3) hold true when the conductor operates at room temperature. The resistance value of resistance–temperature-dependent (RTD) elements varies according to the temperature of the element. The resistance behavior of the materials with temperature can be graphically represented (Fig. 1.18). Especially useful is to represent the resistance behavior of Nickel, Copper, and Platinum with temperature.

The ambient temperature is the temperature at which electrical energy is transported and the conductor achieves thermal equilibrium. The resistivity of the wire at a temperature other than 20 °C is given as

$$\rho_f = \rho_{20} \cdot (1 + \alpha \cdot \Delta T) \tag{1.10}$$

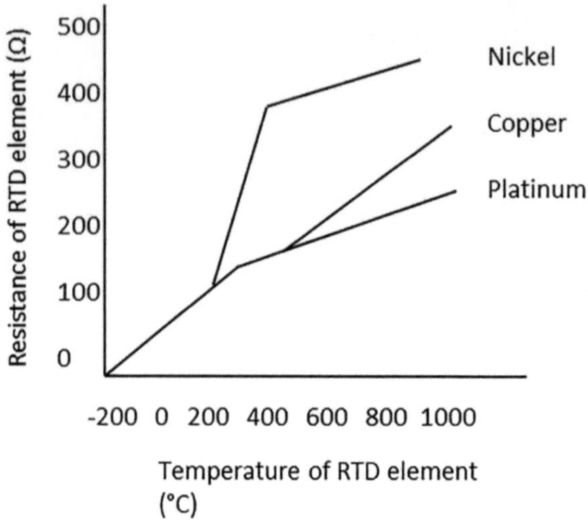

Fig. 1.18 Electrical resistance–temperature curves. *Source* Own elaboration

where ρ_{20} is the conductor resistivity at 20 °C in $\Omega \cdot mm^2/m$, α is the temperature coefficient in $^{\circ}C^{-1}$, ΔT is the difference $\left(T_f - 20\right)$ in °C, and ρ_f is the final resistivity at the final temperature T_f in $\Omega \cdot mm^2/m$.

Just like the resistivity, the resistance reaches a final value of resistance, Rf, when the working temperature changes.

We express R_f as

$$R_f = R_{20} \cdot \left[1 + \alpha \cdot \left(T_f - 20\right)\right] \tag{1.11}$$

where R_f is the final resistance at final temperature, T_f, in Ω, R_{20} is the conductor resistance at 20 °C in Ω, α is the temperature coefficient in $^{\circ}C^{-1}$, and T_f is the final temperature in °C.

This claim is valid if there is a direct correlation between temperature and resistance change. In other words, we assume it only if the resistance remains a linear function between the temperature range of 0 and about 180 °C. The most typical scenarios fall within this range of temperatures. If we know the resistance at 20 °C and at the final temperature T_f, (Eq. 1.11) can be used to find the temperature rise of a conductor. The temperature coefficient of a conductor can also be obtained. If the resistance variations at a given temperature were known, this coefficient could be obtained. The temperature coefficients for various materials are given in table below (Table 1.4). For most metals, the temperature coefficient is positive.

As a result, resistivity and therefore resistance, increases with temperature. On the contrary, other materials such as carbon, germanium, etc. have a negative temperature coefficient. As a result, the resistance of these materials decreases with temperature increases.

Table 1.4 Temperature coefficient of the most commonly used conductors in electrical installations

Material	Temperature coefficient α (°C^{-1})	Resistivity
Copper	0.00392	$1.7 \cdot 10^{-8}$
Aluminum	0,00,403	$2.8 \cdot 10^{-8}$
Nickel	0,006	$7.8 \cdot 10^{-8}$
Almelec	0,00,360	

Source [10]

Example 1.1 Consider a DC motor whose field coils have 310 loops of copper wire with an area of 1.5 mm^2 each. The average length of each turn is 0.6 m, the number of field coils is 4, and they are series-connected. Determine.

Overall resistance at 20 °C.

After 45 min of operation, the winding resistance is measured, obtaining a value of 11 Ω. What is the motor temperature at the operating regime?

Data:

$\text{n} = 310$	Number of copper conductor loops.
$1 = 0.6\,\text{m}$	Average length of each loop, in m.
$\text{N} = 4$	Number of exciter coils.
$\rho_{20} = 0.0173\Omega \cdot \text{mm}^2/\text{m}$	Resistivity of Copper at 20 °C, in Ω·mm^2/m.
$R_f = 11\Omega$	Resistance at final temperature, Tf in Ω.
$\alpha = 0.00392\,°\text{C} - 1$	Temperature coefficient of the conductor, in °C^{-1}.

Hints:

$$\text{L} = \text{n} \cdot 1 \cdot \text{N}$$

$$\rho = \frac{R \cdot S}{L} \tag{1.12}$$

ρ Resistivity coefficient (Ω·mm^2/m).
R Resistance in ohms (Ω).
S Conductor area in mm^2
L Conductor length in meters (m)

$$R_f = R_{20} \cdot \left[1 + \alpha \cdot \left(T_f - 20\right)\right] \tag{1.13}$$

where R_f is the final resistance at final temperature T_f in Ω, R_{20} is the conductor resistance at 20 °C in Ω, α is the conductor temperature coefficient in °C^{-1}, T_f is the final temperature in °C, and ΔT is the increase of temperature experienced by a conductor at known resistances at 20 °C and at final temperature,T_f in °C.

Answer:

n = 310 Number of loops of copper wire, in m.
l = 0.6 m Average length of each loop, in m.
N = 4 Number of excitation coils.

Answer:

$$\mathrm{L} = \mathrm{n} \cdot 1 \cdot \mathrm{N} \rightarrow \mathrm{L} = 310\,\text{loops} \cdot 0.6\text{m} \cdot 4\,\text{coils} = 744\,\mathrm{m}$$

Let us determine the material resistance at 20 ºC using Eq. 1.12

$$\rho = \frac{R \cdot S}{L} \rightarrow R_{20} = \rho_{20} \cdot \frac{L}{S} = 0.0173 \cdot \frac{744}{1.5} = 8.58\Omega$$

where:

ρ Resistivity coefficient (Ω·mm²/m).
R Resistance in ohms (Ω).
S Conductor area in mm²
L Conductor length in meters (m).
R_{20} Resistance of copper in ohms (Ω).

Since we know the resistivity coefficient, $\rho_{20} = 0.0173$ (Ω·mm²/m), then using (Eq. 1.14), ΔT can be determined as

$$R_f = R_{20} \cdot \left[1 + \alpha \cdot \left(T_f - 20\right)\right] = R_{20} \cdot [1 + \alpha \cdot (\Delta T)] \rightarrow \Delta T = \frac{\frac{R_f}{R_{20}} - 1}{\alpha} \tag{1.14}$$

Then, substituting numerical values, we obtain:

$$\Delta T = \frac{\frac{R_f}{R_{20}} - 1}{\alpha} = \frac{\frac{11}{8.58} - 1}{0.0392} = 71.95\,^{\circ}\mathrm{C}$$

Rearranging (Eq. 1.15), we have:

$$R_f = R_{20} \cdot \left[1 + \alpha \cdot \left(T_f - 20\right)\right] = R_{20} \cdot [1 + \alpha \cdot (\Delta T)] \rightarrow \Delta T = \frac{\frac{R_f}{R_{20}} - 1}{\alpha} \tag{1.15}$$

$$\Delta T = \frac{\frac{R_f}{R_{20}} - 1}{\alpha} = \frac{\frac{11}{8.58} - 1}{0.0392} = 71.95\,^{\circ}\mathrm{C} \rightarrow T_f = \Delta T + 20 = 91.95\,^{\circ}\mathrm{C}$$

where:

R_f Strength at final temperature T_f in Ω.
R_{20} Conductor resistance at 20 ° C in Ω.
α Conductor temperature coefficient in ° C-1.

T_f Final temperature in ° C.
ΔT Temperature rise of a conductor with known resistances at 20 ºC and at the final temperature, T_f in ºC.

Example 1.2 Find the resistance and the amount of heat generated by a 100 W halogen lamp. Consider the wire is made from 20 cm length, 0.75 mm diameter, and it has a resistivity of 0.87 Ω mm²/m. Consider that it is switched on for 40 min, and that the current it consumes is 16 A.

Substituting the numerical values we calculate:

$$\mathrm{d} = 0.75\mathrm{mm}$$

$$\rho = 0.87\Omega \cdot \mathrm{mm}^2/\mathrm{m}$$

$$\mathrm{l} = 20\,\mathrm{cm}$$

$$t = 40\,\mathrm{min}$$

$$\mathrm{I} = 16\mathrm{A}$$

First, we find the area of the cylindrical wire as,

$$S = \frac{\pi \cdot d^2}{4} = \frac{\pi \cdot 0.75^2}{4} = 0.44\,\mathrm{mm}^2$$

$$R = \rho \cdot \frac{l}{S} = 0.87 \cdot \frac{0.2}{0.44} = 0.39\Omega$$

Second, we calculate the heat that the lamp produces.

$$Q = 0.24 \cdot I^2 \cdot R \cdot t = 0.24 \cdot 16^2 \cdot 0.39 \cdot 2400 = 57507.84\,\mathrm{cal.}$$

$$1\,\mathrm{Jul.} = 0.2388\,\mathrm{Calories}$$

1.8 Summary

This concluding section provides a essential overview of the chapter's key topics.

This introductory chapter provides a broad contextualization of Electrical Engineering. It covers fundamental concepts starting with the force in a magnetic field and the physical constitution of conductive materials like copper. It distinguishes between conductive and insulating materials and introduces the basic principles of electricity generation and electrodynamics.

The first chapter defines the core elements of an electric circuit and explains key concepts such as electric current and current density. A significant portion is dedicated to electrical resistance, delving into its definition, the factors that affect it (resistivity, conductance, conductivity), and its relationship with temperature.

References

1. Wikipedia. Em fields field (physics).svg. Accessed 12 Dec 2023. [Online]. https://commons.wikimedia.org/wiki/File:Em_fields_field_(physics).svg
2. Pasquale. Carelli, Motion of a charge in a magnetic field. Accessed 4 Dec 2023. [Online]. https://commons.wikimedia.org/wiki/File:Parcar.png
3. Wikipedia. Eddy currents—explanation of drag force.svg. Accessed 4 Dec 2023. [Online]. https://commons.wikimedia.org/wiki/File:Eddy_currents_-_explanation_of_drag_force.svg
4. P. Géry, English: Native copper (2008). Accessed 4 Dec 2023. [Online]. https://commons.wikimedia.org/wiki/File:Cuivre_natif_3(USA).jpg
5. D.J. Griffiths, *Introduction to Electrodynamics*, 3rd edn., Reprinted with corrections (Prentice Hall, Upper Saddle River, NJ, 1999)
6. R.A. Serway, *Principles of Physics*, 2nd edn., Saunders golden sunburst series (Saunders College Publ, Fort Worth, 1994)
7. U. et al. English, Accessed 23 Dec 2023. [Online]. https://commons.wikimedia.org/wiki/File:Electrical_symbols_library.svg
8. Wikipedia. Inductive load, A diagram showing electrons scattering off atoms in a resistive medium (2007). Accessed 11 Dec 2023. [Online]. https://commons.wikimedia.org/wiki/File:Electron_flow_in_a_resistive_medium.svg
9. Chemical Rubber Company, *CRC Handbook of Chemistry and Physics: A Ready-Reference Book of Chemical and Physical Data*, 85th edn. (CRC Press, Boca Raton, 2004)
10. D.C. Giancoli, *Physics: Principles with Applications*, 4th edn (Prentice Hall International, Englewood Cliffs, NJ, 1995)

Chapter 2
Electrotechnics

Abstract In this chapter provides an overview of the main law that we use to solve electrical circuit problems, Ohm's law. Then, the chapter analyzes the various approaches to group resistances. It also defines the DC generator, including the final voltage and the internal resistance. For this reason, understanding the behavior of generators is crucial for using electricity in various applications. Finally, the authors provide an outlook based on the coupling of generators, and electric receivers.

Keywords Ohm's law · Resistances · Generators · Power · Grouping resistors · Generator characteristics · Grouping generators · Electric receivers

2.1 Introduction and Objectives

In the previous chapter, it has been presented some of the components that commonly are found in electrical circuits. We have reviewed their basic measurements. It was found that when a potential difference is applied to a resistor current start to flow. The present chapter investigates the potential of grouping components within circuits. The specific features and forms of aforesaid grouping will be the subject of study. Among the studied components:

- Resistances
- DC generators
- Chemical and mechanical receptors interaction.

2.2 Ohm's Law

Ohm's experiment in 1826 showcased that a two- or three-fold increase in the potential difference in an electric circuit led to a corresponding increase in the current.

$$\frac{U}{I} = \frac{2U}{2I} = \frac{3U}{3I} = \cdots = R \tag{2.1}$$

D. Borge-Diez et al., *Fundamentals of Electrical Power Engineering*, Lecture Notes in Electrical Engineering 1540, https://doi.org/10.1007/978-3-032-18904-2_2

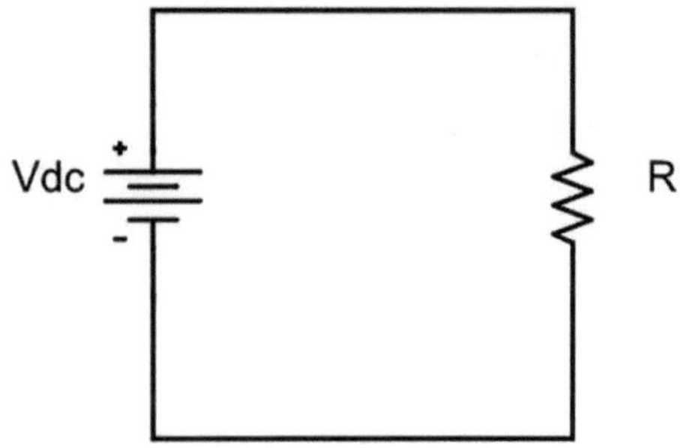

Fig. 2.1 Closed circuit with resistance and voltage source. *Source* Own elaboration

Thus, it follows that the current and potential difference in any conductor will be directly proportional to one another as long as other variables remain constant. In general, the ratio of the voltage V and the current I need not be a constant. This ratio is a constant that we call resistance R, denoting the resistance of the element that follows Ohm's law.

$$I = \frac{U}{R} \tag{2.2}$$

These experiences were embodied in Ohm's law, which states:

Definition The electric current (I) that flows through a conductor between two points is directly proportional to the voltage across the two points (V). The resistance (R) is the proportionality constant and we measure it in ohms (Ω).

The scientific contribution of this law lies in its ability to generalize to any variety of electrodynamics circuit.

Question Which elements are not taken into account in the lumped circuit abstraction?

The following will not be considered in this case: the electromagnetic effects and the emf induction in the conductor because of external magnetic fields. Ohm's law applies to both DC and AC circuits and the elements of a circuit have to verify Ohm's law (Fig. 2.1).

The potential difference is directly proportional to the current. During this chapter we will find an answer to questions like.

What is the scientific contribution of this law?

Which elements are not taken into account?

What are the circuits that will be used for?

2.2.1 *Power Dissipated in Resistance*

Energy is the scalar quantity that characterizes a physical system's ability to do work. When a positive electric charge, q, in an electric field that moves between two potential points, A and B, it gives up energy (does work), w, which is given by the expression:

$$w = q \cdot (v_A - v_B)$$

The Joule is the unit that we use to express energy (symbol J).

We define power as the time derivative of the energy transferred, transformed, or the work. In other words:

$$p = \frac{dw}{dt} = \frac{dw}{dt} = \frac{dq}{dt} \cdot u_{AB}$$

Then;

$$p = u_{AB} \cdot i \tag{2.3}$$

In the case of a linear resistor, we have that:
$u_{AB} = R \cdot i$; Then:

$$P = R \times i^2 \tag{2.4}$$

The power unit is the watt (W).

2.3 Grouping Resistances

Multiple resistors in a circuit can be connected in series, parallel, or a combination of the two.

2.3.1 Grouping Series

Several resistors are connected in series when the current, It, that passes through them is the same. Because of that, electrons follow a single path and the current remains constant throughout the circuit (Fig. 2.2).

$$I_t = \frac{V}{R_T}$$

$$R_T = R_1 + R_2 + R_3 + \cdots \tag{2.5}$$

We can determine the current in the circuit (Fig. 2.2) by Ohm's law, which follows that:

$$U_T = I_T \cdot R_T \tag{2.6}$$

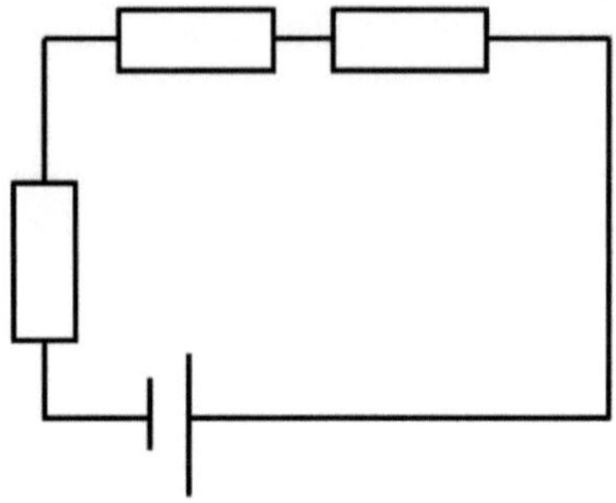

Fig. 2.2 Three series-connected circuit with voltage source. *Source* Own elaboration

The potential difference across each resistor is:

$$U_1 = I_1 \cdot R_1 \tag{2.7}$$

$$U_2 = I_2 \cdot R_2 \tag{2.8}$$

$$U_3 = I_3 \cdot R_3 \tag{2.9}$$

Example 2.1 Consider three resistors with values of 10, 20, and 30 ohms, respectively that are series-connected to a voltage source of 70 V (Fig. 2.3). Please find the electrical current flowing through the resistors, and:

1. Total resistance;
2. Current flowing through the resistors;
3. Voltage across each resistor;
4. Power dissipated by each resistor and
5. Energy dissipated by coupling resistors for 2 h.

Answer:

As they are series-connected, the total resistance is: $Rt = R_1 + R_2 + R_3 = 10 + 20 + 30 = 60\,\Omega$.

According to Ohm's Law, the current flowing through the circuit:

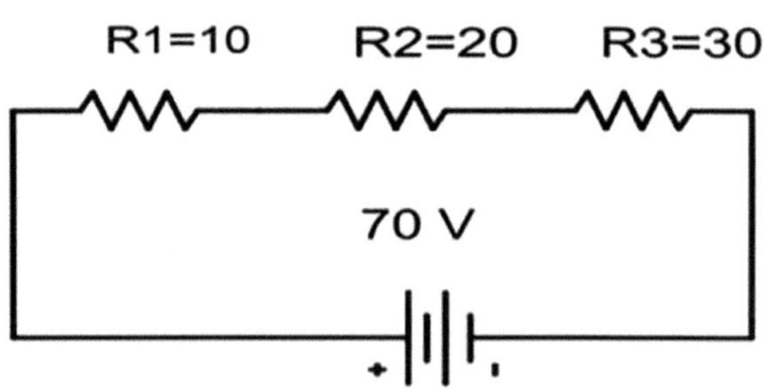

Fig. 2.3 Example 2.1. *Source* Own elaboration

$$I_t = \frac{V}{R_T} = \frac{70}{(10 + 20 + 30)} = 1,1667(A)$$

Since they are series-connected, the current I is the same for the three resistors. Recalling Ohm's law, the potential difference across each resistor's terminals is:

$$V = R \times I$$

$$V_1 = R_1 \times I_1 = 10 \times 1,16667 = 11,6667(V)$$

$$V_2 = R_2 \times I_2 = 20 \times 1,16667 = 23,3333(V)$$

$$\mathrm{V}_3 = \mathrm{R}_3 \times \mathrm{I}_3 = 30 \times 1,16667 = 35(V)$$

The power dissipated by each resistor is:

$$P = R \times I^2$$

$$P_1 = 10 \times 1,16667^2 = 1,36111\,(W)$$

$$P_2 = 20 \times 1,16667^2 = 2,7222\,(W)$$

$$P_3 = 30 \times 1,16667^2 = 4,08\,(W)$$

Then, the total power yields:

$$P_T = \mathrm{P1} + \mathrm{P2} + \mathrm{P3} = 8,1666W$$

The energy dissipated in 2 h:

$$E = 8,1666\,W2\,h = 16,333\,Wh$$

Example 2.2 Consider the circuit below (Fig. 2.4). The battery supplies current to the low-resistance coils in the primary. The transistor opens the primary circuit. When the secondary coil has more loops than the primary coil, it becomes a transformer that increases the voltage. According to Faraday's law, a sudden change in magnetic field causes an increase in voltage.

The primary circuit of a single coil is comprised of 150 loops of copper wire, of 0.6 mm diameter, and a resistance of 0.0172 $\Omega/\mathrm{mm}^2/\mathrm{m}$. The winding has an average diameter of 40 mm and is linked to a 12.6 V source. The remaining part of the primary circuit's resistance is 2 Ω.

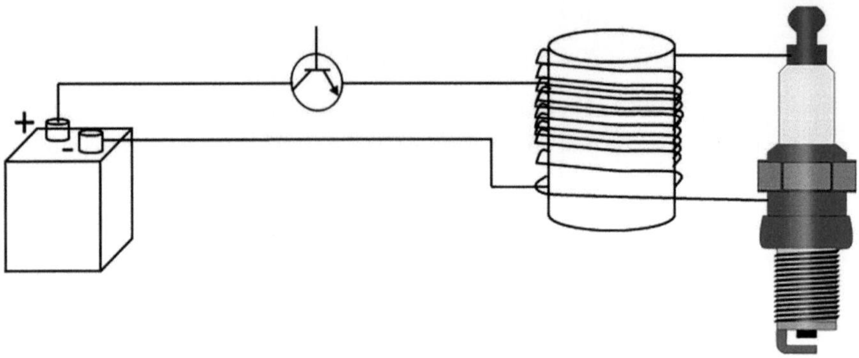

Fig. 2.4 Example 2.2. Series element circuit. *Source* Own elaboration

To determine the heat that the coil generates during one hour with carelessly left contact, firstly find the circuit's total resistance.

We do not know the resistance value of the coil, but we know that the value of the remaining primary circuit is 2 Ω.

First, we have to determine the coil's length and the wire's cross-sectional area within it.

$$L = 2 \cdot \pi \cdot R \cdot N^{\circ} loops$$

$$2 \cdot \pi \cdot 20 \cdot 150 = 18849.55\,\text{mm} = 18.85\,\text{m}$$

$$S = \frac{\pi \cdot d^2}{4} = 0,28\,\text{mm}^2$$

Then, we find the value of the resistance of the coil as:

$$R = \rho l = 0.017218.85 = 1.15\,\Omega$$

The total resistance, as both are series-connected, equals the sum.

$$R_T = R_1 + R_2 = 2 + 1.15 = 3.15\,\Omega$$

Then, applying Ohm's law, we find that the current flowing through the loads is:

$$I_t = \frac{V}{R_T} = \frac{12.6}{3.15} = 4(A)$$

The heat that will be generated is:

$$Q = 0.24 \cdot I^2 \cdot R \cdot t$$

$$Q = 43545.6\,\text{cal}.$$

2.3.2 Grouping Parallel

When some of the current that the voltage source supplies passes through multiple resistors, we say they are parallel-connected resistors. The terminals of the parallel-connected resistors are joint together (Fig. 2.5).

$$I = \frac{V}{R_{eq}}$$

$$I_1 = \frac{V}{R_1}(A)$$

$$I_2 = \frac{V}{R_2}(A)$$

$$I_3 = \frac{V}{R_3}(A)$$

$$I_t = I_1 + I_2 + I_3 \tag{2.10}$$

$$\frac{V}{R_{eq}} = \frac{V}{R_1} + \frac{V}{R_2} + \frac{V}{R_3} \tag{2.11}$$

And dividing it by V:

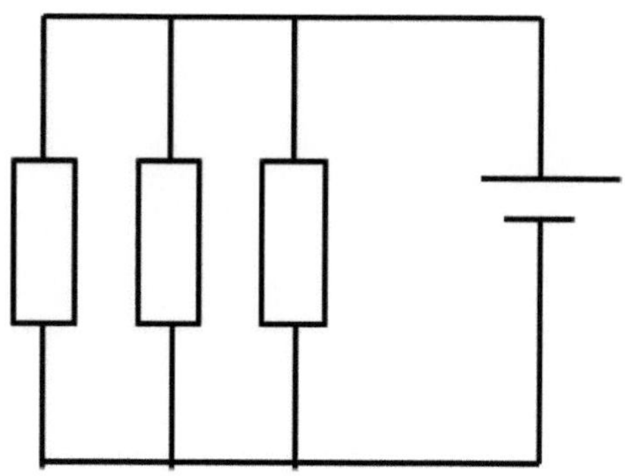

Fig. 2.5 Parallel-connected resistors. *Source* Own elaboration

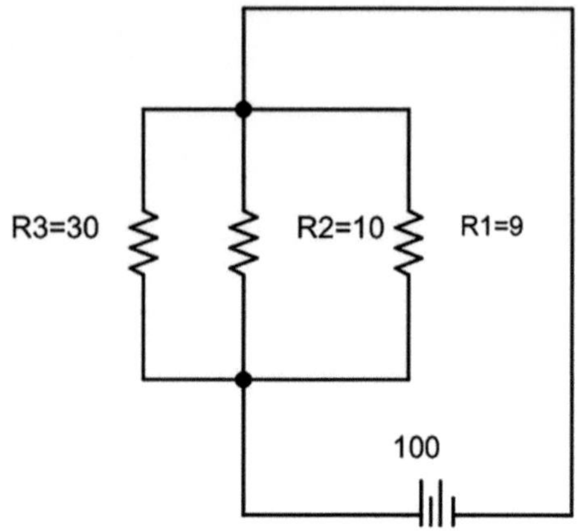

Fig. 2.6 Example 2.3. Parallel-connected resistors. *Source* Own elaboration

$$\frac{1}{R_{eq}} = \frac{1}{R_1} + \frac{1}{R_2} + \frac{1}{R_3} \tag{2.12}$$

Example 2.3 According to the figure below Fig. 2.6, a 100 V voltage source connects to three parallel-connected resistors with values of 9, 10, and 30 ohms. The circuit illustrates a basic parallel resistor network commonly used for calculating equivalent resistance and current distribution.

Determine:

1. The equivalent resistance of the circuit;
2. The current supplied and the current through each resistor.
3. The power dissipated in R_1, R_2, and R_3.

Answer:

Using simple algebra, we find that the equivalent resistance of three parallel-connected resistors is given by:

$$\frac{V}{R_{eq}} = \frac{V}{R_1} + \frac{V}{R_2} + \frac{V}{R_3}$$

$$\frac{1}{R_{eq}} = \frac{1}{9} + \frac{1}{10} + \frac{1}{30}$$

$$R_{eq} = \frac{1}{\frac{1}{9} + \frac{1}{10} + \frac{1}{30}}$$

$$R_{eq} = \frac{1}{\frac{10+9+30}{90}} = \frac{90}{22} = 4.09\ \Omega$$

We find the total current by adding the current in each branch.

$$I_1 = \frac{V}{R_1} = \frac{100}{9} = 11.11\,(A)$$

$$I_2 = \frac{V}{R_2} = \frac{100}{10} = 10\,(A)$$

$$I_3 = \frac{V}{R_3} = \frac{100}{30} = 3.333\,(A)$$

$I_t = I_1 + I_2 + I_3;$
$I_t = 11.11 + 10 + 3.333 = 24.444\,(A);$
According to Eq. 2.13, the power that each resistor consumes is:

$$P_1 = U \times I_1 = R_1 \times I_1^2(W) \tag{2.13}$$

$P_1 = 100 \times 11.11 = 1111(W)$; Or:

$$P_1 = 9 \times 11.11^2 = 1111(W)$$

Power dissipated in R_2:

$$P_2 = U \times I_2 = R_2 \times I_2^2;$$

$$P_2 = 100 \times 10 = 1000(W);$$

Or:

$$P_2 = 10 \times 10^2 = 1000(W);$$

Power dissipated in R_3:

$$P_3 = U \times I_3 = R_3 \times I_3^2;$$

$$P_3 = 100 \times 3.33 = 333(W);$$

$$P_3 = 100 \times 3.33^2 = 1108.89(W)$$

Example 2.4 The figure below Fig. 2.7 shows a circuit with a voltmeter reading 70 V.

In branch 1, we find a resistor made of a material with a resistivity coefficient of 0.85. The resistor is 1000 mm in length, 2 mm in width, and 0.1 mm in thickness.

In the second branch, there is a windscreen wiper motor generating 120 W of power. The third branch comprises 200 loops of copper wire coil, with a radius of

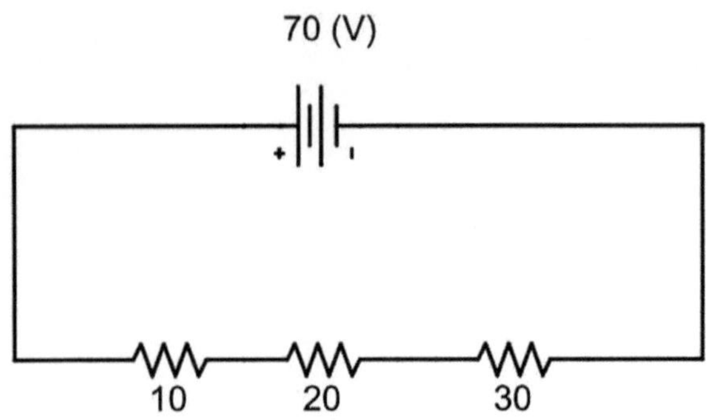

Fig. 2.7 Example 2.4. Resistors in series. *Source* Own elaboration

15 mm and a copper coil resistance of 0.0172 for a conductor with a diameter of 0.4 mm.

Determine:

1. The total resistance of the equivalent circuit;
2. The current in each branch, and
3. The reading of the ammeter.

Answer:

Firstly, we find the equivalent resistance of the circuit.

Let us determine the resistance of each branch. We know that the resistivity of the material in branch 1 is ρ = 0.85. Its length is 1000 mm, its width is 2 mm, and is 0.1 mm thick. From branch 2, we see that the motor has a power of 120 W with a 12.7 V voltage source.

$$P_1 = U \times I_1 = 120\,(W)$$

$$I_1 = \frac{U_1}{P_1} = (A)$$

As we know the voltage and current values, we can derive the value of resistance as:

$$R_2 = \frac{U_1}{I_1} = (A)$$

To answer question 3, let us determine the length and cross-sectional area of the wire, using the data we have available. From these values, along with the resistivity of the material, we can then find the resistance of this branch. We calculate the perimeter of the loop to find the total length.

$$1 = 2\pi \text{ R Number of loops} = 2\pi\ 0.015200 = 18.85\,\text{m}$$

Then, we find the cross-sectional area of the wire as:

$$S = \frac{\pi \cdot d^2}{4} = 706.85\,\text{mm}^2$$

Finally, the coil's resistance is:

$$R = \rho \cdot l = 0.0172 \cdot 18.85 = 0.324\,\Omega$$

$$R_2 = \frac{\rho \times L}{S} = (\Omega)$$

Now, we find the equivalent value of the three parallel-connected resistors.

$$\frac{1}{R_{eq}} = \frac{1}{9} + \frac{1}{10} + \frac{1}{30} = 1.35(\Omega)$$

Substituting these values in the Ohm's law for each component, we find the current in each of the branches:

$$I_1 = \frac{V}{R_1} = (A)$$

$$I_2 = \frac{V}{R_2} = \frac{12.7}{4.25} = 2.98(A)$$

$$I_3 = \frac{V}{R_3} = \frac{12.7}{1.34} = 9.47(A)$$

$$I_3 = \frac{V}{R_3} = \frac{12.7}{2.70} = 4.7(A)$$

Now we find the total current using either Kirchhoff's first law or the equivalent resistance.

$$I_t = I_1 + I_2 + I_3$$

$$I_T = \frac{V}{R_{eq}} = (A).$$

2.3.3 Grouping Resistances: Series–Parallel Circuit

In a series–parallel circuit, the circuit comprises series and parallel resistor combinations between two nodes of the circuit, as shown below Fig. 2.8.

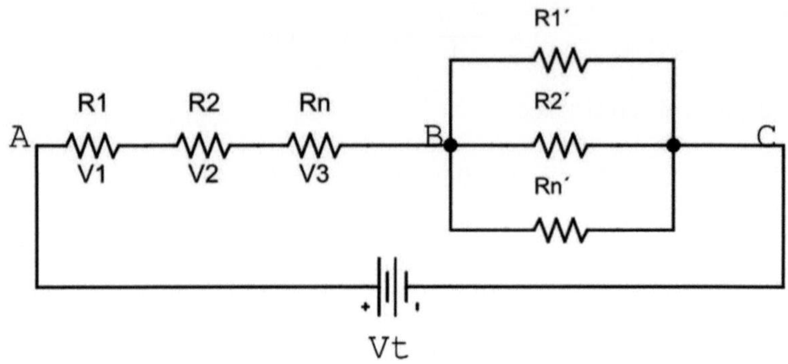

Fig. 2.8 Series–parallel circuits. *Source* Own elaboration

To solve this, we use the equivalent of the parallel-connected resistors between nodes B and C, R_{bc}, and the equivalent of series-connected resistors, between nodes A and B, R_{ab}, and we draw the resulting circuit (Fig. 2.9). Then, we can use Ohm's law to find the voltage across these resistors in the equivalent circuit.

$$U_{AB} = I_T \times R_{AB}$$

$$U_{BC} = I_T \times R_{BC}$$

$$U_{AC} = I_T \times R_{eq}$$

Next, we determine the resistance of the equivalent circuit, and we denote it as R_{eq} or R_t.

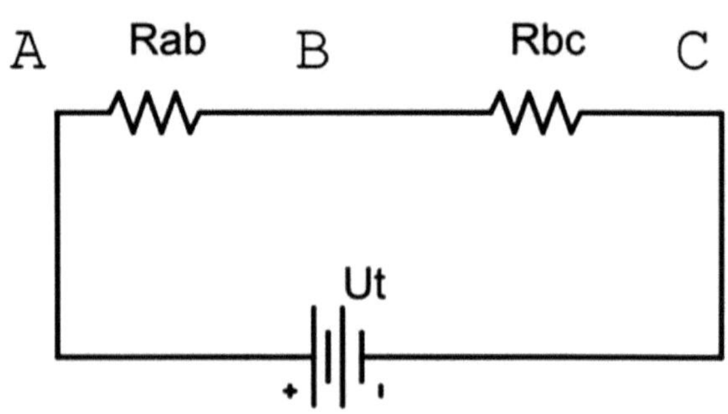

Fig. 2.9 Equivalent series circuits. *Source* Own elaboration

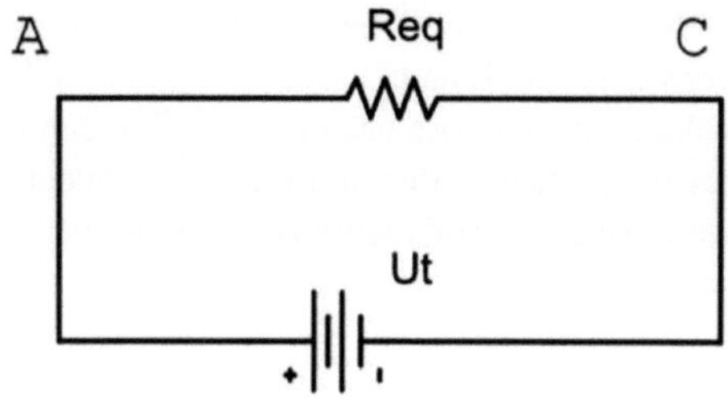

Fig. 2.10 Equivalent reduced circuit. *Source* Own elaboration

$$R_{AB} = R_1 + R_2 + R_n$$

$$R_{BC} = R_{1'} + R_{2'} + R_{n'}$$

Then, the total resistance of the parallel-connected resistors is:

$$\frac{1}{R_{eq}} = \frac{1}{R_{1'}} + \frac{1}{R_{2'}} + \cdots \frac{1}{R_{n'}}$$

Once we determine R_{ab} and R_{bc}, we have an equivalent series circuit, and we can draw the equivalent circuit with R_{eq} as the result of adding R_{ab} and R_{bc}. Figure 2.10.

Since the two resistors are series-connected, the total resistance is:

$R_{AC} = R_{eq} = R_{AB} + R_{BC}$; Then, we can find I_1, I_2, and I_3;

$$I_1 = \frac{U_{BC}}{R_{1'}}$$

$$I_2 = \frac{U_{BC}}{R_{2'}}$$

$$I_3 = \frac{U_{BC}}{R_{3'}}$$

From the reduced circuit (Fig. 2.10), we can use Ohm's law to determine the total current in the circuit. Now, we can find the resulting total current accordingly. Once understood, we can refer back to the previous circuit.

As this is a series circuit, we know the values of the resistors, R_{ab} and R_{bc}. We also know the current passing through them. Recalling Ohm's law, we have:

$$U_{AB} = I_T \times R_{AB};$$

$$U_{BC} = I_T \times R_{BC}$$

Now, we can find the voltage drops between nodes A and B, and the voltage drops between point B and C. According to unreduced circuit (Fig. 2.8), we can deduce the voltage drop across each resistor. We can then find the current flowing through the resistors because they are parallel-connected. Because U_{bc} applies to every branch, we must use Ohm's law to determine the current in each branch.

$$I_{1'} = \frac{U_{BC}}{R_{1'}}$$

$$I_{2'} = \frac{U_{BC}}{R_{2'}}$$

$$I_{3'} = \frac{U_{BC}}{R_{3'}}$$

After solving the mixed circuit, we check the results with Eqs. 2.14 and 2.15:

$$U_1 + U_2 + \cdots + U_n = U_{AB} \quad (2.14)$$

$$U_{AB} + U_{BC} = U_{AC} \quad (2.15)$$

$$I_1' + I_2' + \cdots + I_n' = I_T \quad (2.16)$$

Example 2.5 The voltage source in the series–parallel circuit of figure below Fig. 2.11 is $V_1 = 100$ (V).

Determine:

1. The resistance in each branch;
2. The total resistance;
3. The total current, and
4. The current that flows through each branch.

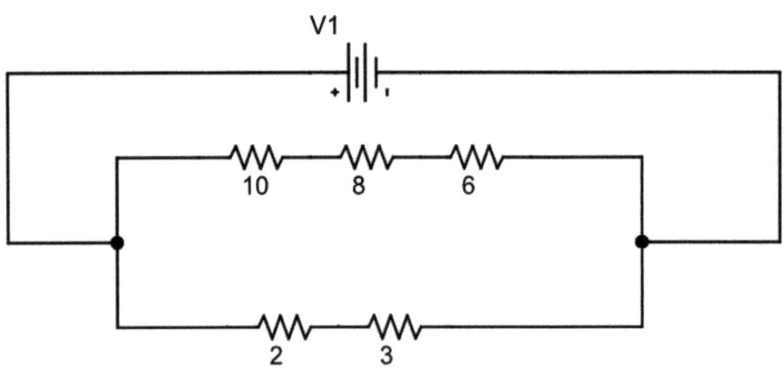

Fig. 2.11 Example 2.5 Series–parallel circuits. *Source* Own elaboration

Answer:
The resistance in each branch:

$$R_1 = 10 + 8 + 6 = 24\,\Omega$$
$$R_2 = 2 + 3 = 5\,\Omega$$

The total resistance is derived from:

$$\frac{1}{R_{eq}} = \frac{1}{R_1} + \frac{1}{R_2}$$
$$\frac{1}{R_{eq}} = \frac{1}{24} + \frac{1}{5}$$
$$R_{eq} = \frac{120}{29} = 4.14\,\Omega$$

According to Ohm's law, the total current is:

$$I_T = \frac{U_T}{R_{eq}} = \frac{100}{4.14} = 24.15(A)$$

Then, the current flowing through each branch is:

$$I_1 = \frac{U_T}{R_{eq}} = \frac{100}{24} = 4.15(A)$$
$$I_2 = \frac{U_T}{R_{eq}} = \frac{100}{5} = 20(A)$$

The total current equals the sum of the branch currents, as can be confirmed.

2.4 DC Generator: Definition

This section will answer questions such as what is an electric generator.

Question How do we define the DC generator?

In electricity generation, a generator converts different types of energy into electrical energy. If a generator delivers alternating current (AC), the polarity changes regularly, following a wave-like pattern. We use different symbols for representing AC generators (Fig. 2.12).

Question What do we define a DC generator?

A DC generator is a device that maintains a constant voltage between its connection points independently of the amount of current drawn from those terminals. The

Fig. 2.12 Symbols representing voltage sources. *Source* From wikicommons under free to use license [1]

current only flows in one direction. We use different symbols to represent a DC voltage sources (Fig. 2.12b, c, d, e).

2.4.1 Power Generator Characteristics

Three main characteristics define a power generator: emf, "V", rated current, "I", and internal resistance, "Ri" which comprise the power generator model (Fig. 2.13). "R_L" models the reduced circuit that connects to the voltage source.

The electromotive force, "V" is measured in volts (V), and sustains the voltage at the generator terminals by propelling free electrons from the negative to the positive terminal in the external circuit. It also delivers electrons from the positive to the negative terminal in the internal circuit of the generator. The figure below (on the left) Fig. 2.14 shows an open-circuit real voltage source, while the figure on the right shows a closed-circuit real voltage source.

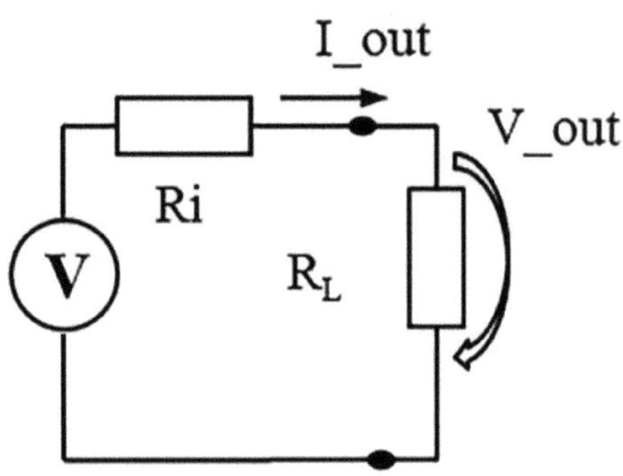

Fig. 2.13 Real voltage source scheme. *Source* Own elaboration

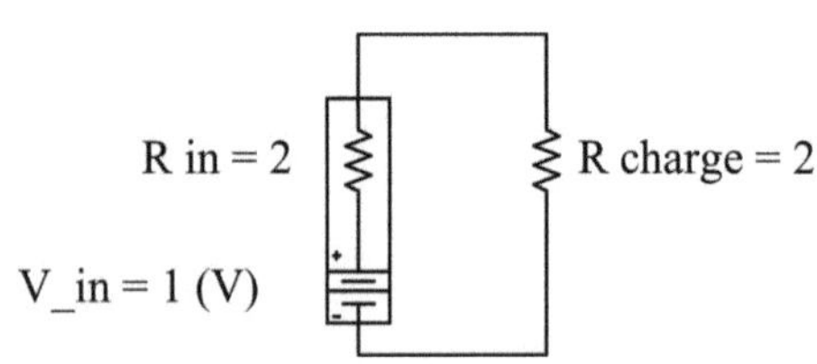

Fig. 2.14 Real voltage sources model. *Source* Own elaboration

2.4.2 Terminal Voltage

The terminal voltage at the generator's poles is the most important quantity of an electric generator. In conclusion, we have:

Ub Terminal voltage (poles), in V
E electromotive force (emf), in V
I Current, in A
R_{in} Internal resistance, in Ω.

2.4.3 Power in a Generator

We will consider the following question: Why the loads cannot use all the power that the source generates?

Question Essentially, what proportion of the generator power is disposable?

The product of the generator's emf and the supplied current is the total output (Pt) that the generator produces. Then, we have:

$$\mathrm{P_t = E \cdot I} \tag{2.17}$$

Question What is the equivalent power loss of the generator?

The power losses in the generator, P_p, is given by:

$$P_p = R_{in} \cdot I^2 \tag{2.18}$$

where R_{in} is the internal resistance of the voltage source and I is the current.

Finally, useful power is the amount of power delivered to the external circuit. The useful power equals the product of the voltage between the nodes and the supplied current:

$$P_u = U_b \cdot I \tag{2.19}$$

Then, we can find Pu, by substituting Eq. 2.5 into (Eq. 2.19):

$$Ub = E{-}r \cdot I \rightarrow Pu = E \cdot I - r \cdot I \rightarrow Pu = Pt - Pp \tag{2.20}$$

Because of the inherent internal resistance, some of the power of the generator dissipates as heat, so not all is available for the connected receivers.

2.4.3.1 Useful Demonstration of Potential

From Kirchhoff's second law, we know that:

$$U_b = E - r \cdot I; P_u = P_t - P_p \tag{2.21}$$

Recalling:

$$U_b = E - r \cdot I$$

Substituting Ub in (2.19) we have:

$$P_u = (E - r \cdot I) \cdot I = E \cdot I - r \cdot I^2 \tag{2.22}$$

Thus, the total power minus the power loss is the useful power. Then, we have:

$$P_u = P_t - P_p. \tag{2.23}$$

2.4.4 Performance of a Generator

This section analyzes the performance of the generator from two viewpoints: firstly, the industrial efficiency, and secondly power consumption. We focus on power generation and power consumption.

Industrial efficiency measures the amount of power, when we compare it to the benefit, taking into account both mechanical and electrochemical alternatives. Obviously, efficiency is always lower than 100%.

We express the efficiency of a generator as:

$$\eta_{ind} = \frac{P_u}{P_{ab}} \tag{2.24}$$

Example 2.6 A 100 V generator has 0.15 Ω of internal resistance, and delivers 10 A of electrical current.

Determine:

1. The voltage at the final terminals of the generator;
2. The generator's total output power;
3. The useful power of the generator, and
4. The power loss in the generator's internal resistance.

Answer:
The final voltage at the terminals of the generator:

$$Ub = E - r\mathrm{I} = 100 - 0.1510 = 98.5\mathrm{V}$$

Total electrical power:

$$Pt = E\mathrm{I} = 10010 = 1000\,\mathrm{W}$$

Useful power of the generator:

$$Pu = Ub\mathrm{I} = 98.510 = 985\,\mathrm{W}$$

The power losses in the generator's internal resistance:

$$Pp = r\mathrm{I}^2 = 0.15 \cdot 10^2 = 15\,\mathrm{W}$$

We can also determine it as:

$$Pp = Pt - Pu = 1000 - 985 = 15\,\mathrm{W}$$

2.5 Connection of Voltage Sources

It is possible to connect several generators to produce varying emf, or different current levels. Depending on the application, we can connect them in series, parallel or a mix of both.

2.5.1 Series Connection of Voltage Sources

Coupling condition

When we connect two or more generators in series, they will connect one after the other. The negative terminal of one generator is then connected to the positive terminal of the next and so on Fig. 2.15. The two unused terminals become the positive and negative terminals of the series connection coupling.

Coupling features: I is common

To prevent overload at current levels above the benchmark, each generator must have the same rated current. The sum of the electromagnetic forces from each generator makes up the total electromagnetic force of the circuit. We express it as:

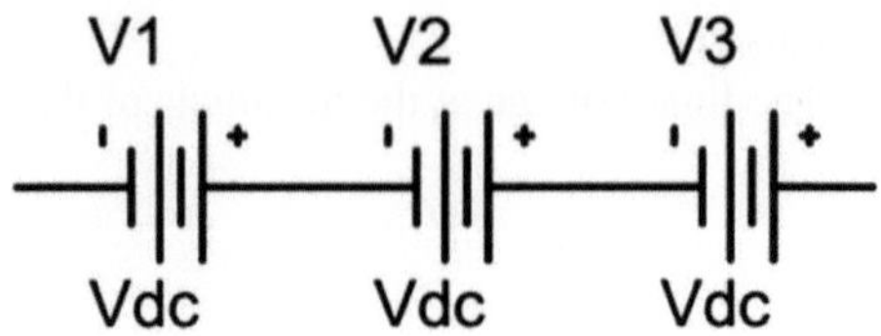

Fig. 2.15 Voltage sources in series. *Source* Own elaboration

$$V_T = V_1 + V_2 + V_3 \tag{2.25}$$

When this condition takes place, the characteristics of the coupling are:

The current is the same for all coupled generators because it is a single path. The addition of the internal resistance of each generator is the coupling's total internal resistance. This is because it is a series connection of internal resistances. Then

$$r = r_1 + r_2 + r_3 \tag{2.26}$$

Question When do we use series connection of voltage sources?

We use series connection of voltage sources when we want to increase the output voltage while maintaining the current equal to that of a single generator.

Why should they have the same insulation?

2.5.2 *Parallel Connection of Voltage Sources*

In this section, we explore the coupling condition and the essential prerequisites to ensure that all generators produce consistent current and power output. We perform a three voltage sources parallel source by connecting the positive terminals between them and similarly for the negative ones (Fig. 2.16).

When several generators are connected in parallel, the positive terminal of a generator connects to the positive terminal of the other generators. This terminal acts as the positive terminal of the parallel coupling. We connect the negative terminals similarly. This kind of coupling requires that the generators have the same emf so

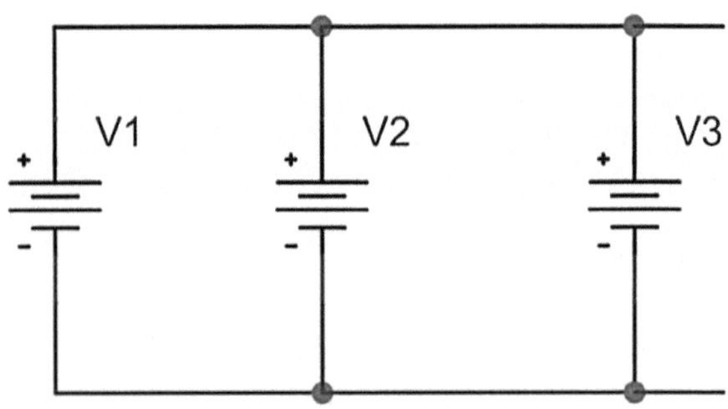

Fig. 2.16 Voltage sources connected in parallel. *Source* Own elaboration

that they can all deliver power. If we want all generators to deliver the same current and power, their internal resistance must also be similar.

Now, we describe the features of the parallel coupling:

The emf of the coupling must be the same for all the generators, and it is the total voltage (Vt).

$$V_t = V_1 = V_2 = \cdots = V_n \tag{2.27}$$

The coupling's total current equals the sum of each generator's current.

$$I_t = I_1 + I_2 + \cdots + I_n \tag{2.28}$$

Finally, the coupling's total internal resistance (rt) equals the inverse of the sum of the inverses of the internal resistances of each generator (internal resistances in parallel).

$$r_t = \frac{1}{\frac{1}{r_1} + \frac{1}{r_2} + \cdots + \frac{1}{r_n}} \tag{2.29}$$

In this instance, the computation is simplified because the internal resistance remains constant across all generators.

Question When do we use a parallel connection?

We connect power generators in parallel to increase the output current while maintaining the voltage equivalent to that of an individual generator.

2.5.3 Series–Parallel Connection of Voltage Sources

The aim of this arrangement is to deliver electric power at a specific voltage and current. Let us now examine the series, and parallel-connected branches. In this section we will examine the series–parallel coupling condition. For the series–parallel connection of six voltage sources we have (Fig. 2.17).

Question When do we perform a mixed coupling?

It results from connecting various groups of generators in series and then linking two or more of these groups in parallel. The emf of both branches must be the same since the equivalent circuit is a parallel circuit. As we have said before, each generator's internal resistance should be similar. By adding the value of the resistors as well of the root mean square (rms) voltage (Eq. 6.47) value in each branch we have:

$$V_1 + V_2 + V_3 = V_4 + V_5 + V_6 \tag{2.30}$$

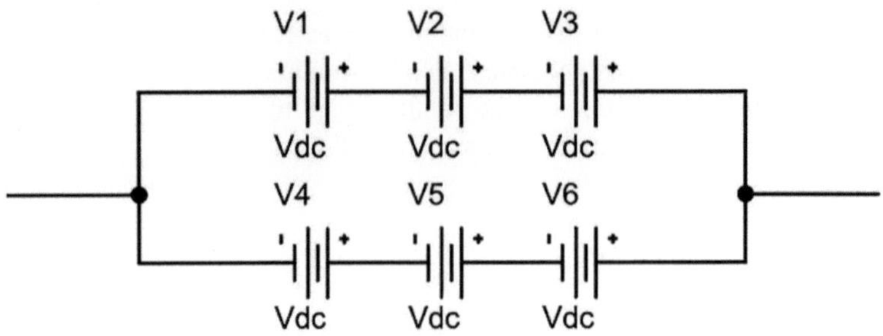

Fig. 2.17 Mixed coupling voltage sources. *Source* Own elaboration

$$r_1 + r_2 + r_3 = r_4 + r_5 + r_6 \tag{2.31}$$

Example 2.7 In the circuit below Fig. 2.18, determine:

1. The total Voltage, Vt;
2. The total resistance, Rt;
3. The total current, It;
4. Ub; and
5. The total Power, Pu.

The voltage drop V_t in the parallel circuit must be the same in both branches. In each branch, the equivalent voltage of the series connection is the sum of the voltage of every generator in each branch.

$$\begin{aligned} V &= \sum Vbranch_1 = \sum Vbranch_2 \\ &= 12 + 12 = 12 + 12 = 24\text{V} \end{aligned}$$

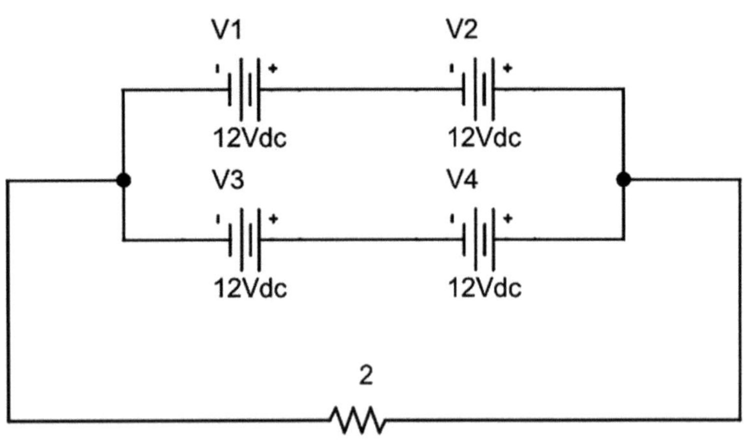

Fig. 2.18 Mixed coupling voltage sources of Problem 2.7. *Source* Own elaboration

Then, the total internal resistance, r_t, is the sum of the series-connected resistances in each branch. Then, the equivalent circuit becomes parallel-connected with another two branches. Therefore, r_t equals the resistance of each series multiplied together, and then divided by the sum of both pairs of resistances in parallel:

$$r_t = \frac{1}{\frac{1}{r_{branch1}} + \frac{1}{r_{branch2}}}$$

In branch 1, we have:
$rbranch_1 = r_1 + r_2$; and in branch 2:
$rbranch_2 = r_3 + r_4$

$$r_t = \frac{1}{\frac{1}{1} + \frac{1}{1}} = 1\Omega$$

To find the total voltage, we subtract the total resistance from the value. The voltage at the terminals is the value minus the voltage drop caused by the circuit's internal resistance. Then, the current is:

$$I_t = \frac{V}{R_t} = \frac{V}{R + r_t} = \frac{24}{2 + 1} = 8A$$

And the voltage is:

$$U_b = V - r_t I_t = 24 - 18A = 16\ \text{V}$$

We can verify this, because it must match the voltage drop that *It* provokes:

$$U_b = I_t R = 2 \cdot 8 = 16\,\text{V}$$

The useful power, P_u, equals the voltage across the terminals multiplied by the current, It.

$$P_u = U_b I_t = 16 \cdot 8 = 128\,\text{W}$$

Finally, we verify this result:

$$P_u = I_t^2 R = 8^2 2 = 128\,\text{W}.$$

2.6 Electric Receiver

So far, we know that the circuit's passive elements mainly comprise capacitors, inductors, and resistors. However, the actual electrical equipment consists of more intricate elements such as transistors, diodes among others. Specifically, disciplines such as electronics, or power electronics comprise other components that are also useful for many applications. It is common that an electric receiver contains various passive elements (Fig. 2.19).

An electric receiver is a device that converts electrical energy into another form of energy. Objectively speaking, an electric receiver is a device whose purpose is to transform electrical energy into heat, mechanical, or others. For instance, a motor comprises resistors, coils, and capacitors, among other components. This is an example of an electrical receiver in action.

Overview

Characteristics

Three factors—emf, rated current and internal resistance—characterize the receiver. Furthermore, this comprises counter-electromotive force (cemf). "E′" stands for the counter-electromotive force, which we express in volts. Similarly, "E" is the electromotive force generated during receiver operation (Fig. 2.20), which opposes the voltage at the terminals of the device. It arises in chemical and electro-mechanical receivers, such as batteries and motors, which can function as generators, or receivers. Resistors, on the other hand, do not generate emf.

Fig. 2.19 Electric receivers. *Source* From wikicommons under free to use license [2]

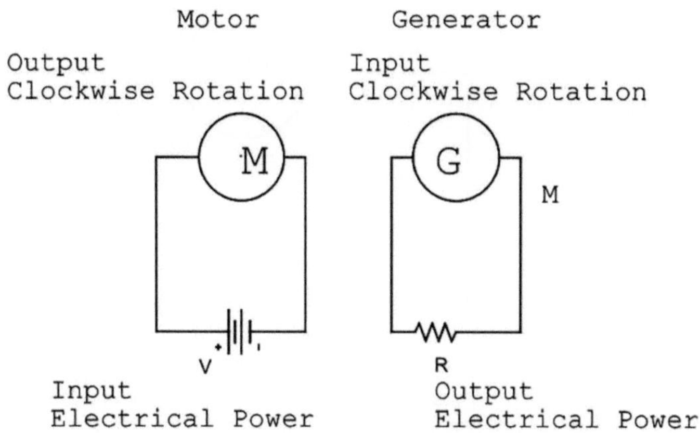

Fig. 2.20 Electric receivers: motor and generator. *Source* Own elaboration

Definition Counter-electromotive force (cemf)

It is the emf that appears while the receiver is working. This force opposes the change in current that induced it.

Question Where does this force come from?

It occurs in chemical (batteries) and electro-mechanical systems (motors) that can act as load or source. This means that they can work as sources or receivers, although cemf does not occur in resistors.

Definition Rated current, In.

Rated current is the maximum current that can flow through a receiver without damaging it. We call it I_n, and we measure it in amperes (A). Finally, the internal resistance, r, represents the resistance to the current flow. Consider a receiver which is part of an electric circuit that absorbs a current I and generates cemf (Fig. 2.21).

Question What will be the electrical voltage at its terminals, V?

According to Kirchhoff's second law, we have:

$$V = E' + r \cdot I \tag{2.32}$$

Fig. 2.21 Electric receiver motor. *Source* Own elaboration

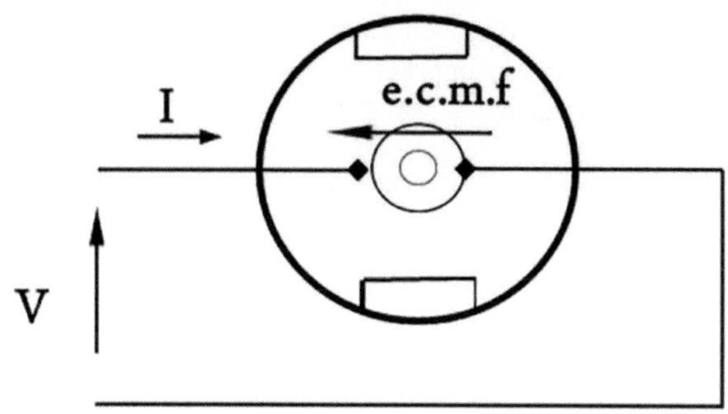

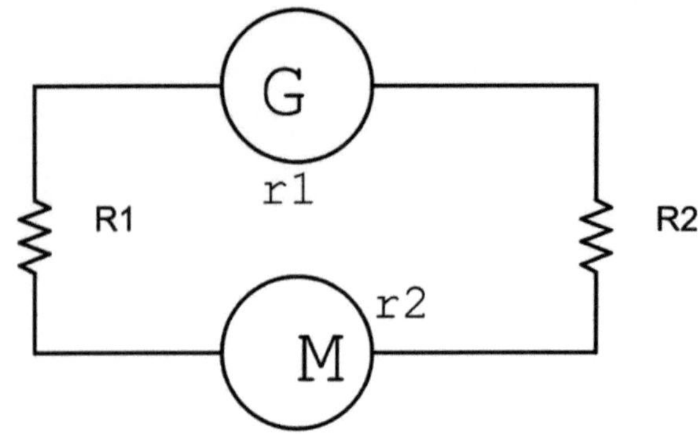

Fig. 2.22 Electric circuit with generator and receiver. *Source* Own elaboration

Question What is dissipated power?

It is the power that the receiver dissipates. It is the product of the voltage across the terminals and the current:

$$P_{ab} = V \cdot I \tag{2.33}$$

The industrial efficiency is the ratio of the useful power that the receiver emits to the power it dissipates.

$$\eta = \frac{P_u}{P_{ab}} = \frac{P_u}{V \cdot I} \tag{2.34}$$

We define an electrical circuit as a set of generators and receivers to form a closed-circuit using wires.

Imagine we have a circuit containing two electrical receivers: a generator and a motor (Fig. 2.22). The internal resistance of the generator and the motor are respectively r_1 and r_2. It also contains two resistances R_1 and R_2.

Let us apply Kirchhoff's second law to find the equation of the branch:

$$E_G - E'_M = I \times (r_1 + r_2 + R_1 + R_2) \tag{2.35}$$

What does the cemf means regarding the emf?

In a circuit with a generator and a receiver (chemical or electro-mechanical), the emf of the receiver opposes to that of the generator.

Then, we have:

$$E - E' = I \cdot (r_1 + r_2 + R_1 + R_2) \tag{2.36}$$

To find the current that will flow through our circuit, we clear I:

Question What is the current that flows through the circuit?

$$I = \frac{E - E'}{r_1 + r_2 + R_1 + R_2} \tag{2.37}$$

Example 2.8 Find the current that flows through a circuit that contains two 45 W head lamps each when we apply a voltage of 12 V to terminals.

Answer:

As there are two headlights, the total power is 90 (W). This is the current that will pass through our circuit under normal conditions:

$$I = \frac{P}{U} = \frac{90}{12} = 7.5A$$

2.7 Simulating Circuits with LTspice

LTspice Project Workflow (Quick Summary).

2.7.1 *Installation*

Download LTspice from the Analog Devices official website.

Install it on Windows or macOS.

On macOS, drag LTspice into the Applications folder.

2.7.2 *Starting a New Project*

First, we open LTspice. Go to: → File → New Schematic. Then,

Blank canvas appears for circuit design. Now it is time for placing components. The main shortcuts we have are:

R → resistor

C → capacitor

L → inductor

G → ground (mandatory)

If we strike F2 (Windows)/fn + F2 (Mac) it appears the component library (sources, transistors, etc.). These must be connected with wires (F3). For editing values, we just have to Right-click (or Ctrl + click on Mac) a component to set values. Examples are: 10 k for resistor, 5 V for voltage source; and units: k (kilo), Meg (mega), m (milli), u (micro), n (nano), p (pico).

2.7.3 *Simulation*

Once we have drawn the circuit it is time for adding simulation commands. First, we must to insert a SPICE directive (S key). Among the common commands we have:.op → DC operating point. .tran 1 m → transient analysis (1 ms). .ac dec 10 1 1 k → AC sweep. Then, we have to place directive text box on schematic. Now, we are ready for running the simulation. Just Click Run icon (man running). Then LTspice executes analysis and opens results window.

Click wires → voltage plots.

Click components → current plots.

2.7.4 *Viewing Results*

Numerical results: View → SPICE Error Log.

Graphical results: waveform viewer.

Example 2.9 This example illustrates the simulation results of three basic circuits that we have previously reviewed. We solve them by using LTspice, using the commands that we have presented above (Figs. 2.23, 2.24, 2.25, 2.26, 2.27 and 2.28).

We see the line illustrating the twelve volts constant voltage, and the amperes the voltage source delivers, it is eight three. In addition, we see the amperes that the resistors consume: four, two, and two three.

We see the line illustrating the twelve volts constant voltage, and the amperes the voltage source delivers, it is eight three. In addition, we see the amperes that the resistors consume: four, two, and two three.

We see the line illustrating the twelve volts constant voltage, and below the amperes the resistor two consumes: two mill amperes. Below, the amperes that resistor three consumes: zero. five mill amperes. Below, the voltage in node three, eight. 1 V. Below, the voltage in node two: six volts. Then the amperes through resistor four: minus one point one millamperes. Then the voltage in node two: two point one volts, and finally, the current that the source delivers, it is almost minus two mill amperes.

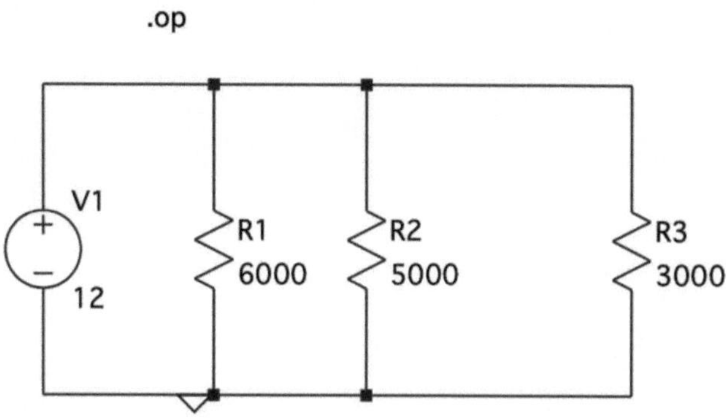

Fig. 2.23 Electric circuit with generator and three parallel-connected resistors. *Source* Own elaboration

Fig. 2.24 Simulation results plot. *Source* Own elaboration

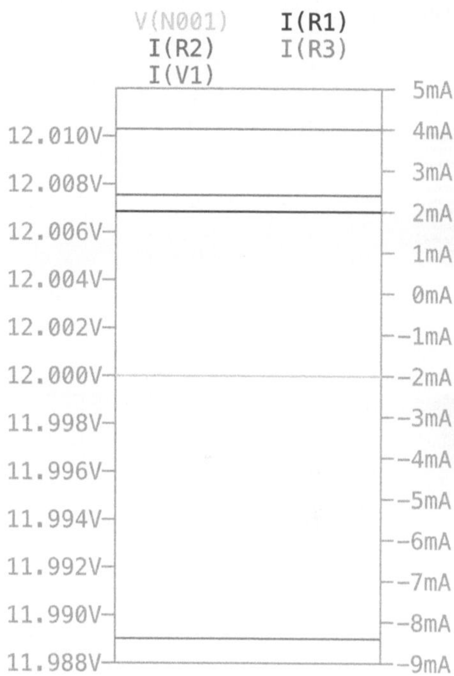

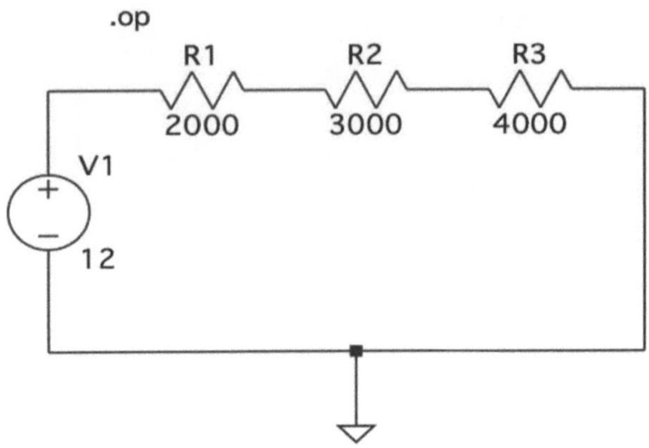

Fig. 2.25 Electric circuit with generator and three series-connected resistors. *Source* Own elaboration

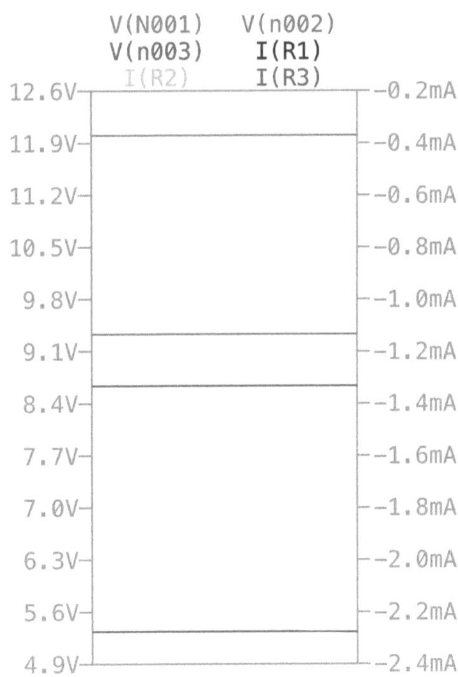

Fig. 2.26 Simulation results plot. *Source* Own elaboration

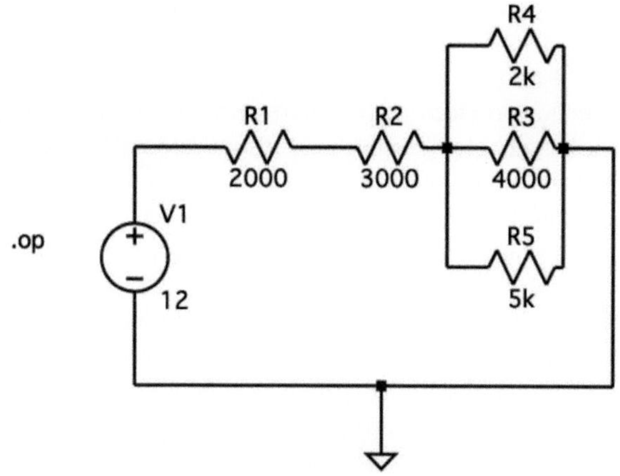

Fig. 2.27 Electric circuit with generator and mixed circuit of resistors

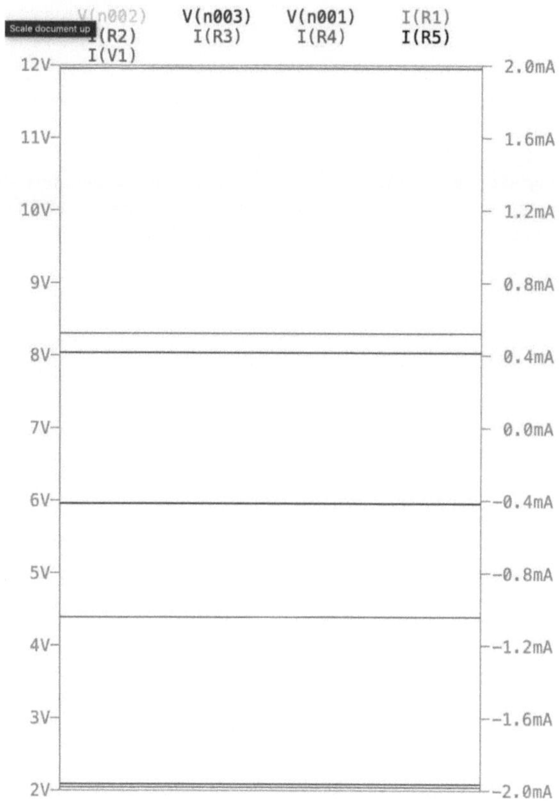

Fig. 2.28 Simulation results plot. *Source* Own elaboration

2.8 Summary

This concluding section provides a essential overview of the chapter's key topics. This chapter, titled "Electrotechnics," establishes the foundational laws and components of basic electrical circuits. It introduces Ohm's Law and the concept of power dissipated in a resistance. A key focus is on the various methods for grouping resistances, covering the rules and calculations for series, parallel, and mixed series–parallel circuits.

The chapter also defines a DC generator, explaining its characteristics, terminal voltage, power output, and performance. It details how to connect these voltage sources in series, parallel, and series–parallel configurations. Finally, it introduces the concept of an electric receiver as the component that consumes electrical energy in a circuit.

References

1. U. et al., English. Accessed 23 Dec 2023. [Online]. https://commons.wikimedia.org/wiki/File:Electrical_symbols_library.svg
2. R. electrician, English: Revox B215 cassette-IO board. The board has been subject to at least two rounds of capacitor replacement, and now it's on the bench again. This time, the playback-critical CMOS switches will be replaced with relays, too. Hope it works. 2022. Accessed 11 Dec 2023. [Online]. https://commons.wikimedia.org/wiki/File:Revox_B215_-_IO_board_-_as-is_(with_many_prior_fixes).jpg

Chapter 3
Capacitance and Inductance: Behavior in DC Circuits

Abstract In this chapter discusses the key characteristics of the passive components in an electric circuit. As we mentioned, we will focus on the behavior of resistors, capacitors, and inductances when we place them in DC circuits. This chapter examines the behavior and characteristics of these components. In addition, we will describe the grouping conditions. Finally, this chapter provides an outlook on the R-L, and R-C circuits.

Keywords Capacitors · Capacitor characteristics · Capacitor behavior · Grouping capacitors · Coils · Coils characteristics · Coils behavior · Grouping coils · Conductors and isolators

3.1 Introduction. Dimensions and Operation

In addition to resistors, electronic circuits include other components such as capacitors and coils. These show different behavior when we place them in circuits with direct current sources. A typical configuration for a capacitor contains two parallel and conductive plates separated by a gap of length d (Fig. 3.1). We observe the dielectric between the conductive plates.

This section discusses the nature, associated magnitudes, and function of capacitors and coils when we place them in resistor networks powered by DC generators.

3.2 Capacitors

Question What is a capacitor?

A capacitor is a component that can store electrical charge or electrons. When we place a capacitor in a circuit we represent it using two parallel lines (Fig. 3.2).

D. Borge-Diez et al., *Fundamentals of Electrical Power Engineering*, Lecture Notes in Electrical Engineering 1540, https://doi.org/10.1007/978-3-032-18904-2_3

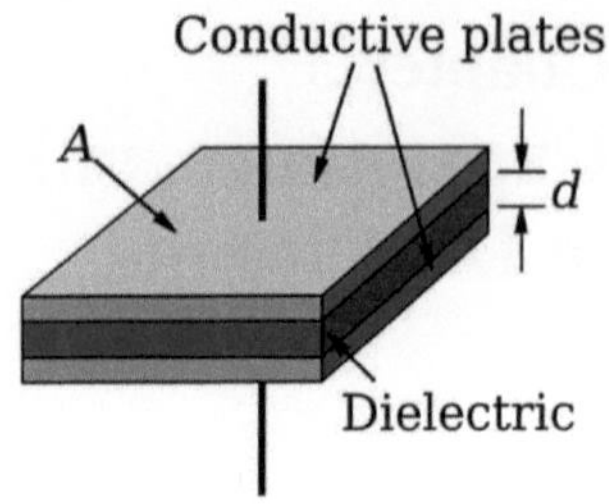

Fig. 3.1 A parallel-plate capacitor. *Source* From wikicommons under free to use license [1]

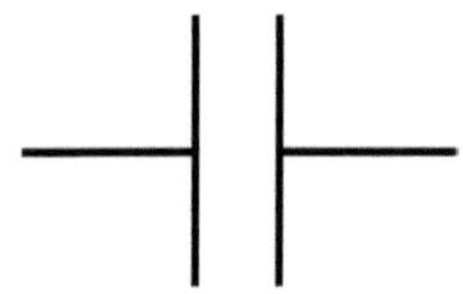

Fig. 3.2 Capacitor symbol. *Source* Own elaboration

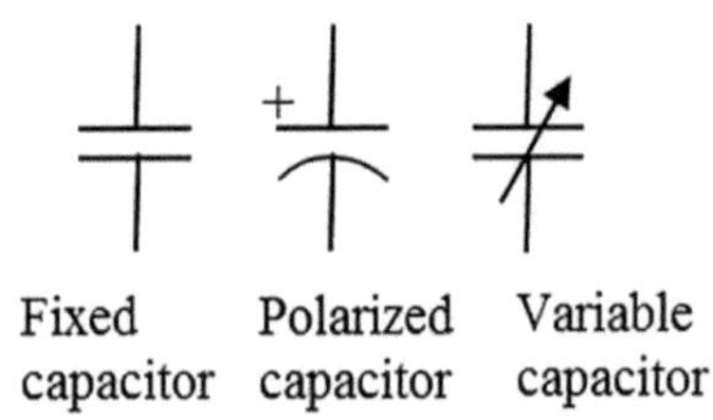

Fig. 3.3 Schematic representations of different types of capacitors. *Source* Own elaboration

We also use other used symbols (Fig. 3.3) for capacitors with enhanced functionalities.

Question What are the main components of a capacitor?

A capacitor consists of two metallic conductive surfaces, called armatures, which are separated by an insulating linear dielectric having a permittivity, ε. It is a measure of the ability of a material to interact with an electric field and become polarized by the field, so that the field within the material is weaker than the field in absence of the material.

3.2.1 Basis of Capacitor Action

Once a capacitor is totally charged, the two conductors will have equal charges but with opposite signs in each terminal. We use the mutual capacitance model of the capacitor (Fig. 3.4).

It is important to keep in mind that a capacitor stores a specific amount of charge and sustains a voltage level, but does not allow direct current to pass through it. Furthermore, it is worth noting that, for calculation purposes, the charge Q of a capacitor belongs only to one of its plates. Charges equally distribute across the

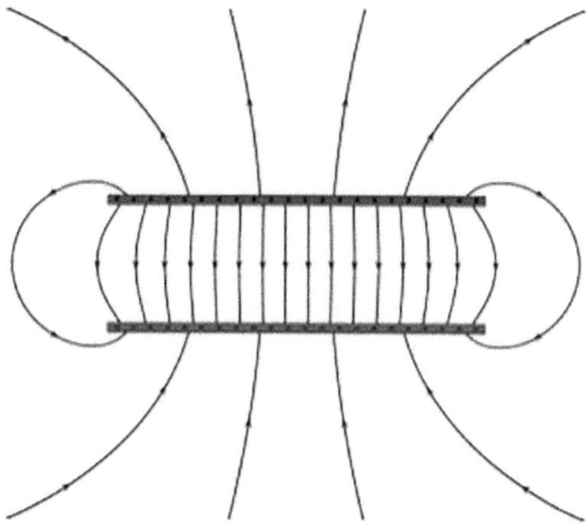

Fig. 3.4 Mutual capacitance model of a capacitor. *Original Source* From wikicommons under CC 2.0 use license [2]

capacitor plates (Fig. 3.5). The greater the area, the higher the charge the capacitor can hold.

> **Important**
> Then, we can draw the following conclusion:
> Once the capacitor reaches the voltage, no current flows through the circuit. We can model the current entering the plates from the starting point until the capacitor reaches the total charge (Fig. 3.6). The voltage across the plates and the current into the plates vary as a capacitor charges.

Having this into account, it is clear that a capacitor inhibits the flow of direct current. According to this observation, a capacitor blocks the flow of direct current. A charged capacitor keeps its potential difference even after disconnection from the charging source.

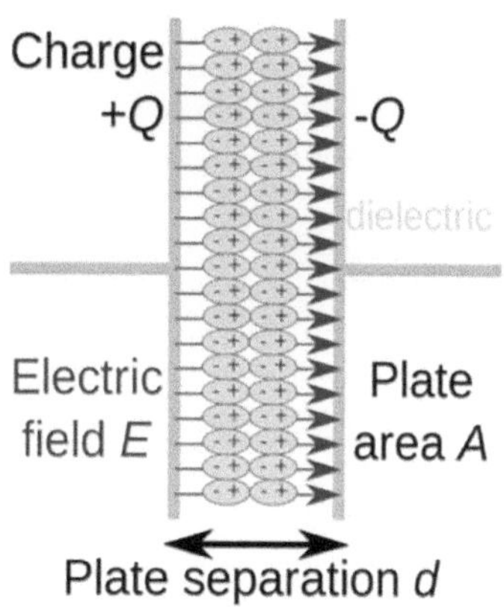

Fig. 3.5 Charges in a capacitor. *Source* From wikicommons under free to use license [3]

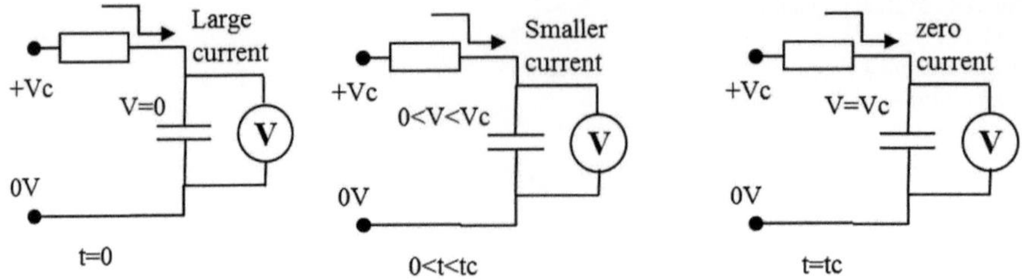

Fig. 3.6 Voltage stage across the plates. *Source* Own elaboration

3.3 Function of a Capacitor

Among the many applications of a capacitor, it is used in electronic circuits to remove DC current from an electrical signal, or to store voltage for later release. The relationship between charge, the capacity of a capacitor and the applied voltage are closely bound (Fig. 3.7).

3.3.1 Capacity of a Capacitor

As we know, a capacitor stores electrical charges. The capacitance refers to the measure of the storage capacity. Specifically, it is the ratio of the electric charge transferred from one to the other of a pair of conductors to the resulting potential difference between them. The capacity depends on the metallic surface where charges are stored, the distance between them, and the permittivity of the medium between the plates (Fig. 3.8).

$$Q = CV \tag{3.1}$$

$$E = \frac{V}{d} \tag{3.2}$$

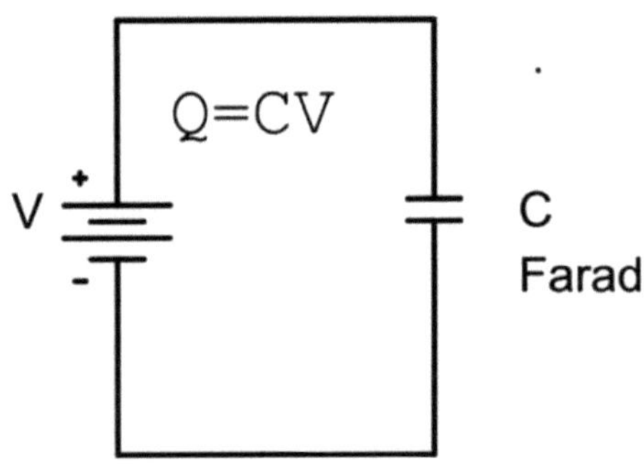

Fig. 3.7 Capacitor charge, and voltage. *Source* Own elaboration

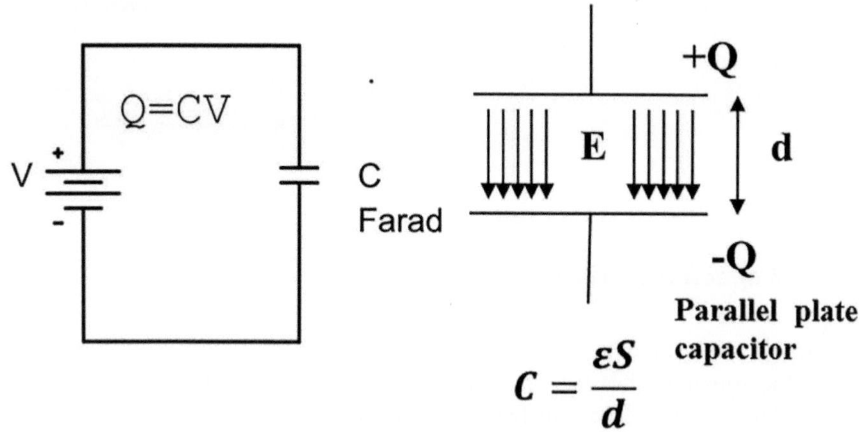

Fig. 3.8 Relationship between the capacitor and the voltage. *Source* Own elaboration

$$C = \frac{\varepsilon S}{d} = \frac{\varepsilon_0 \varepsilon_r S}{d} \tag{3.3}$$

where ε_0 is the permittivity of the vacuum (8.85418782 1012), ε_r is the permittivity of the material, and S denotes the surface of the plate.

We define the capacitance of a capacitor as the ratio of the charge Q that it accumulates on one of its plates to the voltage between its terminals. The farad (F) is the unit of capacitance, which is equal to the capacitance of a capacitor that stores the charge of 1 C with a potential difference of 1 V between its plates (Table 3.1).

Because the farad is a very large unit, we commonly use millifarad, microfarad, nanofarad and picofarad.

Table 3.1 The Farad scale

Yotta Farads	10^{24} (YF)	Farads 1(F)	Deci Farads	10^{-1} (dF)
Zetta Farads	10^{21} (ZF)		Centi Farads	10^{-2} (cF)
Exa Farads	10^{18} (EF)		Mili Farads	10^{-3} (mF)
Peta Farads	10^{15} (PF)		Micro Farads	10^{-6} (μF)
Tera Farads	10^{12} (TF)		Nano Farads	10^{-9} (nF)
Giga Farads	10^{9} (GF)		Pico Farads	10^{-12} (pF)
Meg Farads	10^{6} (MF)		Femto	10^{-15} (fF)
Kilo Farads	10^{3} (kF)		Atto	10^{-18} (aF)
HectoFarads	10^{2} (hF)		Zepto	10^{-21} (zF)
Deca Farads	10^{1} (daF)		Yocto	10^{-24} (yF)

Source Own elaboration

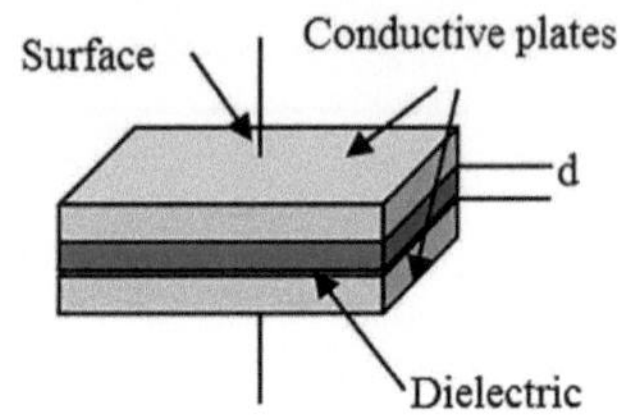

Fig. 3.9 Geometry of a capacitor. *Source* Own elaboration

3.3.2 Capacitance Value of a Capacitor

We know that the amount of charge that a capacitor can store relies on the geometry area of the armatures, but also on the properties of the insulating material between them, and its thickness. The main features of a capacitor are the area of the conductive plates, the distance between the plates, and the dielectric (Fig. 3.9).

We can present it using equivalent and parallel planar structures. We define

$$C = \frac{\varepsilon_0 \cdot \varepsilon_r \cdot S}{d} \tag{3.4}$$

$$C = \frac{8.8510^{-12} \cdot \varepsilon_r \cdot S}{d}$$

C is the capacitance of the capacitor, having the units of Coulombs/Volt, or Farads. S is the surface area in square meters. The larger the area of the armature, the more electricity it can accumulate. D denotes the thickness of the dielectric in meters. The shorter the distance between the armatures, D results in higher charge accumulation.

The permittivity in vacuum, denoted as ε_0 has a value of 8.85 10^{-12} F/m, while we represent the relative permittivity of dielectric substances as ε_r.

$$k = \varepsilon_r = \frac{\varepsilon_m}{\varepsilon_0} \tag{3.5}$$

where ε_m is the material permittivity, and k is the relative permittivity. We have to multiply ε_0 by ε_r to obtain the epsilon value for different substances. It is useful to know the range of ε_r values for various dielectric substances (Table 3.2).

3.3.3 Characteristic Quantities of a Capacitor

When we use a capacitor for a circuit it is important to consider other values beyond the capacitance, although it is the most important characteristic of this component. The operating voltage, the maximum voltage that the capacitor's terminals can tolerate, or the temperatures are factors to consider. The working voltage of a capacitor refers to the maximum voltage that it can handle without suffering damage.

Table 3.2 ε_r values for some dielectric substances

Material	Relative permittivity ε_r
Vacuum	1.0000
Air	1.0006
PTFE, FEP (Teflon)	2.0
Polypropylene	2.20 to 2.28
Polystyrene	2.4 to 3.2
Wood (Oak)	3.3
Bakelite	3.5 to 6.0
Wood (Maple)	4.4
Glass	4.9 to 7.5
Wood (Birch)	5.2
Glass-Bonded Mica	6.3 to 9.3
Porcelain, Steatite	6.5

Source [4]

The working voltage is commonly shown on the surface plate of the capacitor (Fig. 3.10). It also shows the working temperature and the capacity of this capacitor. This consideration highlights the importance of the working voltage in circuit design with capacitors. The working voltage has a significant impact on the size of the capacitor. If we define

$$C = \frac{\varepsilon S}{d} = \frac{\varepsilon_0 \varepsilon_r}{d}, \text{ and } Q = CV$$

Fig. 3.10 Capacitor characteristics. *Source* From wikicommons under free to use license [5]

Fig. 3.11 Burnt high-voltage capacitor of type LYSL6KV5D on an adaptor-inverter board of type AIVP-0065C. *Source* From wikicommons under free to use license [6]

Note that capacitors with higher working voltages must have a greater thickness of the dielectric (d in the denominator). To increase capacitance in these cases, the area (S) (numerator of the expression) must increase significantly.

3.3.3.1 Maximum Voltage of a Capacitor

When a capacitor withstands high voltage, the resulting current can cause the dielectric to break down and puncture it Fig. 3.11. Let us now overlook how the distance of the elements and the increasing voltage between them increases the attraction between the charges of the two armatures.

Question What is the maximum voltage that a capacitor can withstand without piercing its dielectric?

The maximum voltage that a capacitor can withstand without piercing its dielectric is therefore referred to as the 'voltage rating' of the capacitor.

3.4 Grouping Capacitors

What is the aim of grouping several capacitors?

Among the reasons to group capacitors, we highlight to achieve the required capacitance or the desired working voltage by connecting several capacitors together.

3.4.1 Capacitor Series

When we connect capacitors in series, we say it is a series configuration. We have a series connection of capacitors when connect the terminals as a string (Fig. 3.12).

In this case, the overall voltage at the final terminals corresponds to the sum across the ends of each capacitor. Then we have:

$$U = U_{C1} + U_{C2} + U_{C3} \tag{3.6}$$

All capacitors charge equally. Positive charges on C_1 generate corresponding negative charges on the other end, which then balance with positive charges on C_2 and so on.

Each capacitor, in turn, causes a voltage drop. The capacitors have the same charge:

$$Q = Q_1 = Q_2 = Q_3 \tag{3.7}$$

Remember that, by definition:

$$C = \frac{Q}{U} \rightarrow U_i = \frac{Q_i}{C_i}$$

We have:

$$U = U_1 + U_2 + U_3 = \frac{Q}{C} = \frac{Q_1}{C_1} + \frac{Q_2}{C_2} + \frac{Q_3}{C_3} \tag{3.8}$$

Recalling (Eq. 3.7)

$$Q = Q_1 = Q_2 = Q_3$$

Dividing by Q yields:

$$\frac{1}{C} = \frac{1}{C_1} + \frac{1}{C_2} + \frac{1}{C_3} \tag{3.9}$$

We can say that for a series connection of capacitors:

$$Q = Q_1 = Q_2 = Q_3$$

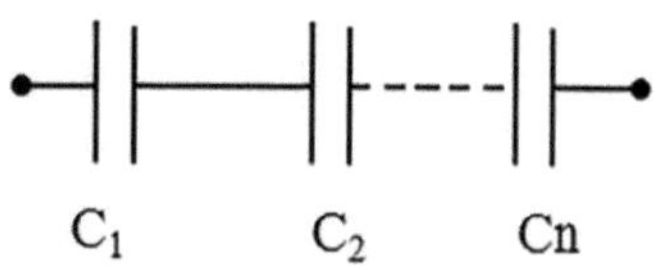

Fig. 3.12 Series-connected capacitors. *Source* Own elaboration

$$U = U_1 + U_2 + U_3$$

$$\frac{1}{C} = \frac{1}{C_1} + \frac{1}{C_2} + \frac{1}{C_3}$$

3.4.2 Capacitor Parallel

When we connect capacitors in parallel, the voltage at the ends of the coupling is equal to the voltage between the ends of each capacitor. Each capacitor acquires a charge according to its capacitance. In this case, the total charge of the coupling equals the sum of the charges of each capacitor. If we connect the two terminals between them we have a group of parallel-connected capacitors (Fig. 3.13).

Because:

$$C = \frac{Q}{U}$$

We have that:

$$Q = C \cdot U_T$$

$$Q_1 = C_1 \cdot U_1$$

$$Q_2 = C_2 \cdot U_2$$

$$Q_3 = C_3 \cdot U_3$$

So that:

$$C \cdot U_T = C_1 \cdot U_1 + C_2 \cdot U_2 + C_3 \cdot U_3 \tag{3.10}$$

And how:

$$U_T = U_1 = U_2 = U_3, \tag{3.11}$$

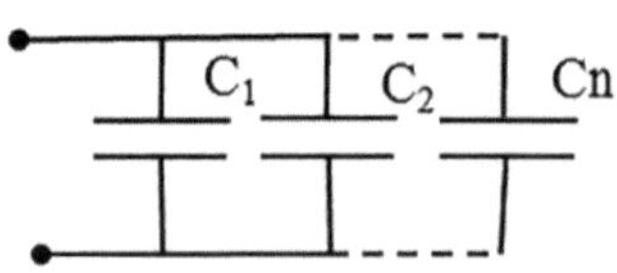

Fig. 3.13 Capacitors grouped in parallel. *Source* Own elaboration

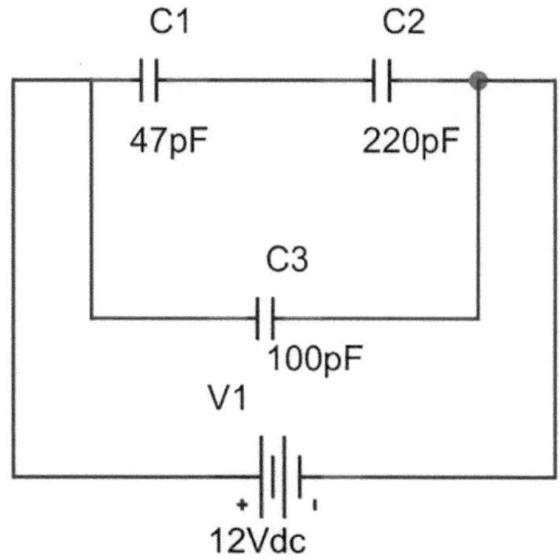

Fig. 3.14 Figure of Problem 3.1. *Source* Own elaboration

Simplifying U, we have that:

$$C = C_1 + C_2 + C_3$$

Problem 3.1 In the circuit below Fig. 3.14 determine:

3.5 Capacitor Behavior

This section reviews the charging progress of a capacitor connected in series to a resistor in a closed circuit. We call this an R–C circuit because it includes a resistor and a capacitor (Fig. 3.15). Closing switch S results in a zero potential difference across the capacitor and the resistor immediately sees the generator voltage. Then, the capacitor starts charging and distributes voltage between both elements for a period of time.

What determines the charging time of a circuit with a capacitor is the moment at which the fully charged capacitor cancels the current flowing through the circuit. At this point, the potential difference across the capacitor equals that of the generator emf, while the voltage drop across the resistor is zero. The values of the resistor and the capacitor determine the time it takes to charge the capacitor. U expresses the charging voltage of the capacitor (Eq. 3.12). We can graph the voltage at the capacitor´s terminals that this equation shows (Fig. 3.15):

$$u = U_0 \cdot \left(1 - e^{-\frac{1}{RC}}\right) \tag{3.12}$$

where U_0 is the Generator voltage, R is the resistance, C is the capacitance, and e is the base of the napierian logarithms whose value is 2.71828.

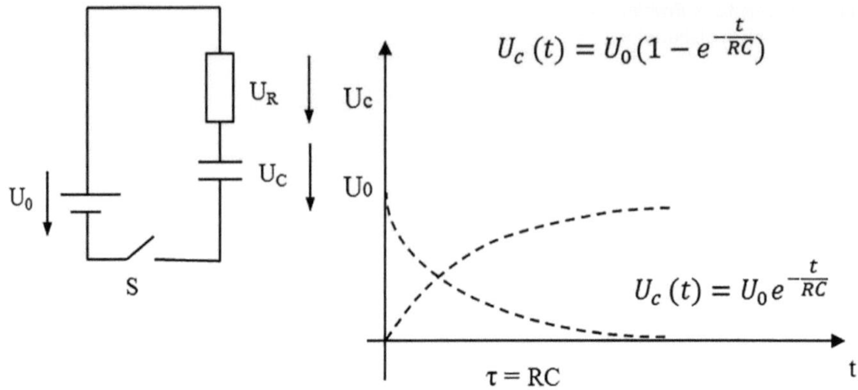

Fig. 3.15 Charging and discharging a capacitor via a resistor. *Source* Own elaboration

Unlike other mentioned quantities, whose value remains constant and written with a capital letter (E, I, R, etc.), we write the potential difference of the capacitor u in lowercase because it is a variable quantity.

A similar process to the charging process occurs when a charged capacitor with a voltage of E connects to a resistor through a switch and discharges. The discharge time relies on the same factors as its charging counterpart. In this case, we find the capacitor's voltage during discharge using the following equation:

$$u = E \cdot e^{-\frac{\tau}{RC}}$$

where we define:

$$\tau = RC \tag{3.13}$$

The time constant is the product of resistance, and capacitance (RC) as we see in the equations above. The time that the capacitor takes to gain 63% of the complete charge during the process, or to release 63% of the full charged capacitor is the same. We also see that the maximum charge and discharge level are attainable only when time reaches infinite (Eq. 3.12).

After five-time constants, the voltage between the capacitor's ends is close to the final value (E or 0). Depending on the process, we consider the charging or discharging process. Capacitors store energy as voltage, because they accumulate charge between its plates, and release it when a consumption circuit is connected, resulting in a net zero-energy consumption process.

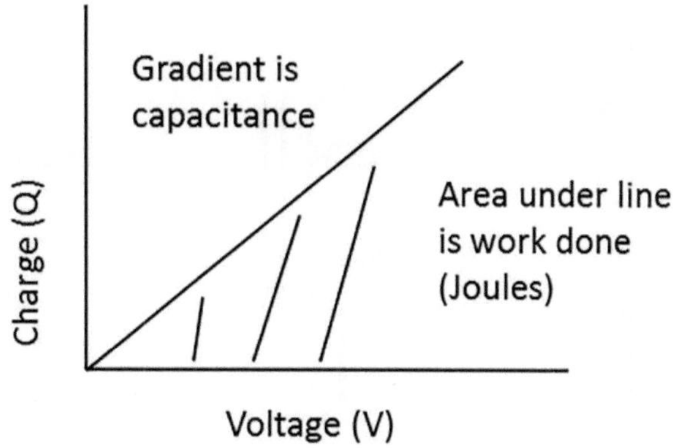

Fig. 3.16 Energy stored in a capacitor. *Source* Own elaboration

3.6 Energy Stored by a Capacitor

Unlike a resistor, a capacitor stores energy rather than dissipates it. Recalling the equation of the capacitor action, we define the energy storage process as:

$$p = v \cdot i = C \cdot v\frac{dv}{dt} \tag{3.14}$$

$$w = \int pdt = C\int v\frac{dv}{dt}dt = C\int_{-\infty}^{t} vdv = \frac{1}{2}C \cdot v^2 \tag{3.15}$$

The process of energy storage in a capacitor follows a linear process (Fig. 3.16).

3.7 Inductances

Besides to resistors and capacitors, we highlight another type of passive element commonly used in circuit abstraction: inductors. We commonly use some symbols to represent inductances (Fig. 3.17).

As we have seen, a capacitor models the effect of the electric field. Correspondingly, the inductance models the effect of magnetic field. Inductances or coils are passive components that consist of a conductive wire that is often wound around a ferromagnetic core material, which is made of magnetic sheet or ferrite.

Question Then how do we define inductance?

Inductance is a property that occurs in an inductor when electric current flows through it, provoking that a magnetic flux appears according to physics laws. In essence, inductance is the number of flux linkages per current unit (amperes).

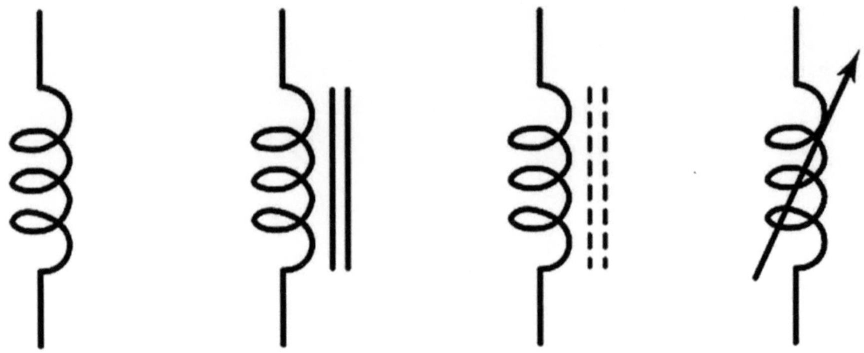

Fig. 3.17 Inductance symbols. *Source* From wikicommons under free to use license [7]

3.7.1 How Do Inductances Store Energy?

When storing energy within a coil, the current flowing through the inductor creates a magnetic field. The energy is stored as a magnetic field. Then we define the energy that the inductor store as

$$E = \frac{1}{2}LI^2 \tag{3.16}$$

If we have an inductance in a circuit with an incandescent light bulb, we will see the delay of the bulb from lighting up because the electrical energy is stored in the magnetic field. Inductors store energy in a magnetic field. We use Henries (H) to find the inductance of an electrical device.

3.7.2 Basis of Inductances Action

We consider two cases in which it is necessary to investigate the behavior of the current when flowing through an inductor. First, when the current flowing through it remains constant, and secondly a circuit with variable current flowing. In a DC circuit, we consider the current steady when it reaches a stable flow after a transient period. The transition into and out of steady state does not occur immediately (Fig. 3.18).

When a circuit connects, it takes a short period to reach the steady state. During the transition interval the current fluctuates from zero to steady-state current (Fig. 3.18). During this transient interval, the coil produces an emf that is opposite in direction and equal in magnitude to the voltage supplying it, unlike the steady state. It is known as Lenz's law, and we define it as:

$$e = -L\frac{di}{dt} \tag{3.17}$$

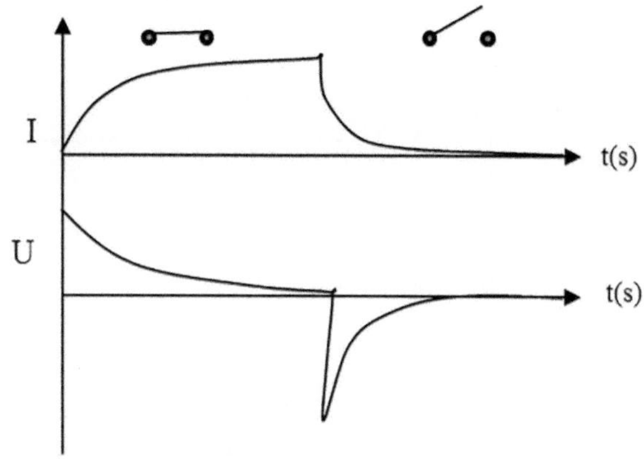

Fig. 3.18 Transient behaviour of current and voltage in an inductance. *Source* Own elaboration

L denotes the self-induction coefficient (in Weber/Ampere, or Henries). Self-inductance is the ratio of the electromotive force produced in a circuit by self-induction to the rate of change of current producing it. The henry is the unit of self-inductance and is the value that a coil needs to generate 1 V of emf when the current changes at a rate of 1 A over a period of one-second. We define the voltage as:

$$V = N\frac{d\phi}{di} \tag{3.18}$$

We know, that flux is: $\phi = \frac{Li}{N}$; Then:

$$V = L\frac{di}{dt} \tag{3.19}$$

If V = 1 V;

$$\frac{di}{dt} = \frac{1\,\text{amp}}{\text{s}}, \text{ then inductance} : L = 1\,\text{H}.$$

3.8 Grouping Inductances

We can connect inductors (coils) in either series or parallel, like resistors, or capacitors.

Fig. 3.19 Inductance series. *Source* Own elaboration

3.8.1 Inductance Series

A bunch of series-connected inductances occurs when we connect the terminal of an inductance to another (Fig. 3.19). Series connection of inductances (coils) increases the total inductance of a circuit. Grouping inductances in series results in a single inductance with a value equal to the sum of the individual inductances:

$$L_T = L_1 + L_2 + L_3 + \cdots + L_n \tag{3.20}$$

$$L_{eq} = \sum_{i=1}^{i=n} L_i$$

3.8.2 Inductance Parallel

We have parallel-connected inductances when both of their terminals connect to each terminal of other inductance. Then, a parallel connection is one that connects the terminals of several inductances among them (Fig. 3.20).

We see that when coils are parallel-connected, we know that i_{L1} plus i_{L2}, and so on, results in i_T. We also know that we define the self-induced emf across an inductor as $L\frac{di}{dt}$. Due to this fact, the following applies:

$$V = L_T\left(\frac{di_1}{dt} + \frac{di_2}{dt} + \frac{di_n}{dt}\cdots\right)$$

Fig. 3.20 Inductance parallel. *Source* Own elaboration

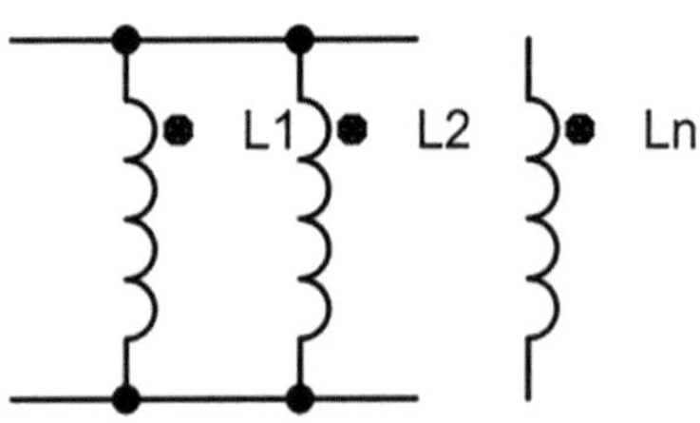

Then, substituting $\frac{di}{dt} = \frac{V}{L}$, the value of U, we obtain:

$$U = U_{\mathrm{L1}} = U_{\mathrm{L2}} = U_{\mathrm{L3}}$$

$$\frac{1}{L} = \frac{1}{L_1} + \frac{1}{L_2} + \frac{1}{L_3}$$

Thus, we see that the reciprocal inductances in parallel results in:

$$\frac{1}{L_T} = \frac{1}{L_1} + \frac{1}{L_2} + \cdots \frac{1}{L_n} \tag{3.21}$$

$$\frac{1}{L_{eq}} = \sum_{i=1}^{i=n} \frac{1}{L_i}$$

3.9 Inductance Behavior

Except for some transient moments, when we directly apply direct current to a coil is similar to a resistor having a resistance equals to its own internal resistance. A circuit containing a switch, a coil, and a voltage source is a R-L circuit (Fig. 3.21).

Let us now examine the transient behavior of an inductor in a DC circuit. This will give us an insight of how alternating current circuits operate.

Recalling $e = \frac{d\Phi}{dt}$, Where Φ represents magnetic flux and "e" denotes the emf generated (lower case because is a variable magnitude). This is how we explain the relationship between the two:

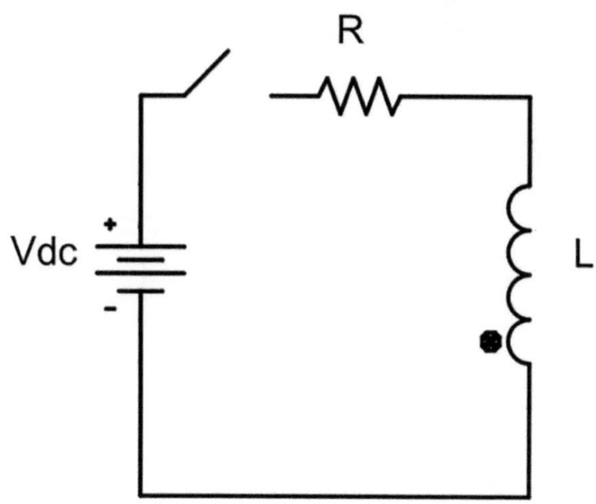

Fig. 3.21 Inductive R-L circuit. *Source* Own elaboration

$$\Phi = L \cdot i \tag{3.22}$$

We have:

$$d\Phi = Ldi \tag{3.23}$$

Since L is constant. The result is:

$$e = -L \cdot \frac{di}{dt} \tag{3.24}$$

Upon closing, a self-induced current appears which opposes the flow of current that generates it.

3.9.1 Current Flowing Through a Circuit That Includes Self-induction and Resistance

It is worth analyzing a network comprising a coil with an inductance value of L Henries, and a resistor fed by a DC generator (Fig. 3.22). When the switch S closes, the self-inductance becomes a generator, and then the emf that appears opposes the current that flow through the circuit. When the switch S closes, the current through this network will be zero due to the opposing generator force known as the counter-electromotive force. It takes some time for the current to reach the steady state value.

The time it takes to reach the steady state depends on the inductance L and resistance R values. The current flow progression in a R-L circuit can be plotted

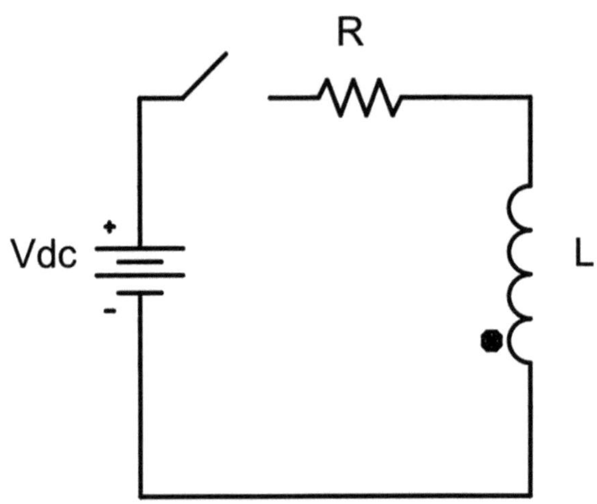

Fig. 3.22 R-L circuit. *Source* Own elaboration

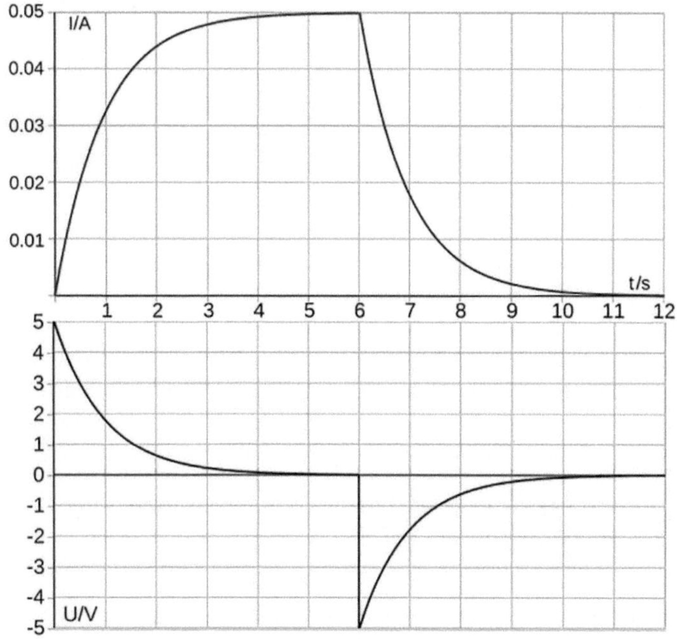

Fig. 3.23 Current flow progression in a RL circuit being: L = 100 H; R = 100 Ω; U_0 = 5 V. *Original Source* From wikicommons under CC 2.0 use license [8]

(Fig. 3.23). It illustrates the behavior of the magnitudes current and voltage in a RL circuit. When t = 0, the current is null and gradually increases to equal E/R when t = infinity. If a generator in an inductive network experiences a short circuit while a current of magnitude E/R is flowing through it, the current does not dissipate immediately but continues to circulate for some time. The R and L values also determine how long this current flows.

The expression that relates the value of the current to self-inductance and resistance is as follows:

$$i = \frac{E}{R} \cdot \left(1 - e^{-\frac{t}{L/R}}\right) \tag{3.25}$$

$$V_L = V_b \cdot \left(e^{-\frac{t}{L/R}}\right) \text{ (Voltage drop across a coil)}$$

Then, we determine the current value:

$$i = \frac{E}{R} \cdot \left(e^{-\frac{t}{L/R}}\right) \tag{3.26}$$

We define the time constant as:

$$\tau = \frac{L}{R} \tag{3.27}$$

The L/R ratio determines the time constant in an inductive circuit. The current either increases or decreases by approximately 63% during the first time constant. After five-time constants (t = 5L/R), the current will achieve its final value. Self-inductance stores energy during the current build-up process that occurs when the switch is closed until its value stabilizes.

Question

What happens with the stored energy once the generator is removed?

Once we remove the generator, the current begins to decrease, releasing the energy into the circuit.

3.9.2 Energy Stored in a Coil

A coil stores energy as a magnetic field, and we measure the energy in joules. We express it as:

$$W = \frac{1}{2} \times LI^2 \tag{3.28}$$

Problem 3.2 Suppose we have a coil with a 100 mH of coefficient of self-induction where a current that varies from 0 to 10 A passes through it during 4 ms.

What is the self-induced emf?

3.10 Summary

This chapter explains the fundamental principles of capacitors and inductors in Direct Current (DC) circuits. It begins by defining what capacitors and inductances are, detailing their construction, operational basis, and key characteristics such as capacitance value and maximum voltage.

The chapter then covers how to group these components, explaining the rules for connecting capacitors and inductors in both series and parallel configurations. Finally, it describes their transient behavior in DC circuits, focusing on how they store and release energy over time, and provides the formulas for calculating the energy stored in both a capacitor's electric field and an inductor's magnetic field.

References

1. Vector drawing of parallel plate capacitor. Accessed 4 Dec 2023. [Online]. https://publicdomainvectors.org/es/vectoriales-gratuitas/Dibujo-vectorial-de-condensador-de-placas-paralelas/26061.html
2. VFPt capacitor-square-plate.svg. Accessed 11 Dec 2023. [Online]. https://commons.wikimedia.org/wiki/File:VFPt_capacitor-square-plate.svg
3. Wikipedia. Capacitor schematic with dielectric.svg. Accessed 4 Dec 2023. [Online]. https://commons.wikimedia.org/wiki/File:Capacitor_schematic_with_dielectric.svg
4. H.D. Young, R.A. Freedman, yH.D. Young, *University Physics with Modern Physics*, 15th edn. (Pearson Higher Education, Hoboken, N.J., 2020)
5. G. Houston, English: S. I. brand capacitor, 1000 μF, 25 V, apparently made by Transfull Limited of Hong Kong (2015). Accessed 4 Dec 2023. [Online]. https://commons.wikimedia.org/wiki/File:S.I.-capacitor-20150807-003.jpg
6. Cmuelle, Deutsch: Defekter Kondensator auf einer Stromversorgungsplatine (adaptor-inverter board) von Lien Chang des Typs AIVP-0065C, welche in einem LCD-Fernseher der Marke LG, Modellbezeichnung M2762D, verbaut wurde. Die Platine wurde in der 34. Woche in 2009 produziert. 2014. Accessed Dec 2023. [Online]. https://commons.wikimedia.org/wiki/File:Burnt_high-voltage_capacitor_of_type_LYSL6KV5D_on_an_adaptor-inverter_board_of_type_AIVP-0065C_by_Lien_Chang_power-supplying_an_M2762D_TV_by_LG.JPG
7. U. et al., English. Accessed 23 Dec 2023. [Online]. https://commons.wikimedia.org/wiki/File:Electrical_symbols_library.svg
8. Frankemann, Norsk bokmål: Inductor switch-on-off: 2015. Accessed 10 Dec 2023. [Online]. https://commons.wikimedia.org/wiki/File:Inductor_switch-on-off_positive_voltage.png

Chapter 4
Behavior of Passive Elements in AC Circuits

Abstract This chapter uses Ohm's law for the analysis of purely resistive or purely capacitive circuits in the sinusoidal steady state. After analyzing a pure inductive circuit, Faraday's law is presented. This chapter finishes with an analysis of the sinusoidal RL, RC, and RLC series circuits.

Keywords Sinusoidal steady state · Ohm's law · Passive elements in alternating current · Faraday's law · RC circuit · RLC circuit

4.1 Introduction. Ohm's Law in Steady State. Sinusoidal

Passive elements such as resistors, coils, and capacitors behave differently in direct current (DC) compared to alternating current (AC). The voltage sources that circuits can contain can be either direct or alternating sources (Fig. 4.1).

While in direct current the polarity remains constant in alternating current the polarity reverses at equal intervals of time. Because alternating current is variable, new phenomena will occur compared to direct current, particularly because of the use of coils and capacitors as receivers in circuits as we have seen in previous chapter. We can draw how a capacitor behaves in both alternating, and direct current (Fig. 4.2).

Question Why does a capacitor block DC but allow AC?

We have to remember that the three passive elements (resistors, capacitors and coils) limit the current in a circuit, as described by Ohm's law for AC:

$$I = \frac{U}{Z} \tag{4.1}$$

Although Ohm's law for alternating current (Eq. 4.1) use rms values, we can also use instantaneous values. Then, we have that:

$$i = \frac{u}{z} \tag{4.2}$$

D. Borge-Diez et al., *Fundamentals of Electrical Power Engineering*, Lecture Notes in Electrical Engineering 1540, https://doi.org/10.1007/978-3-032-18904-2_4

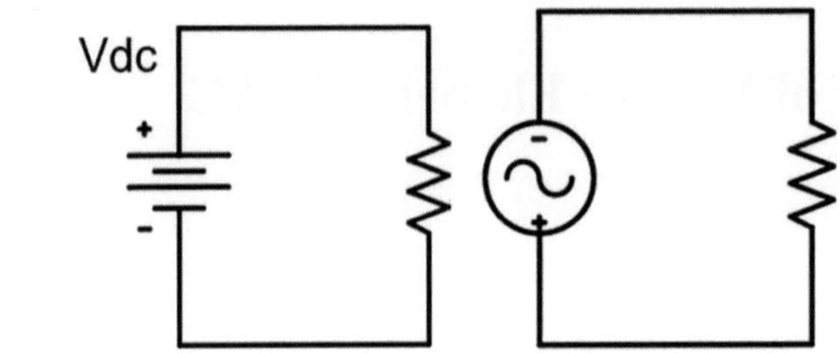

Fig. 4.1 Direct Current voltage source, and sinusoidal voltage source. *Source* Own elaboration

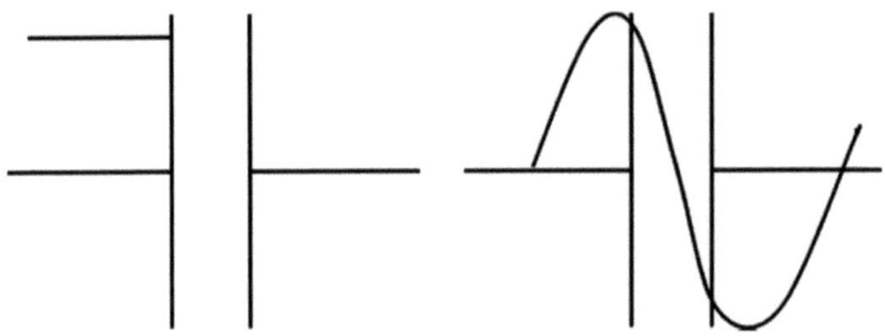

Fig. 4.2 Behavior of capacitors in DC and AC. *Source* Own elaboration

Next, we look at how passive elements, resistors, coils and capacitors behave in alternating current.

4.2 Purely Resistive Circuit

When we apply a sinusoidal voltage source to a resistor, a current appears according to the characteristic or defining equation of the "resistive" element. This current generates heat due to the Joule effect. Slightly different, a circuit that contains a sinusoidal voltage source can also have resistors, coils, capacitors, among others (Fig. 4.3).

Fig. 4.3 Purely resistive circuit. *Source* Own elaboration

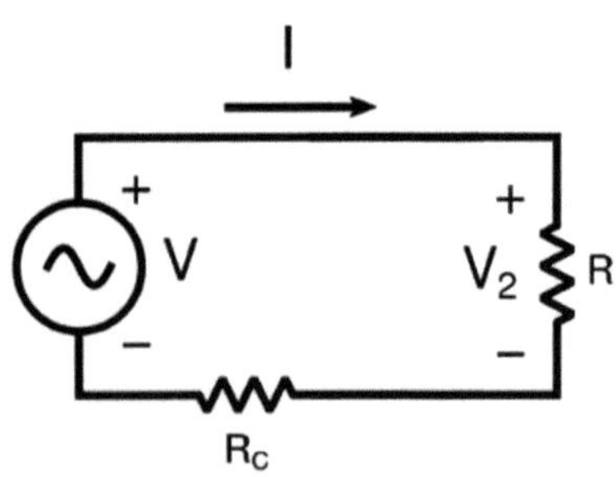

Only by dividing the instantaneous value of the voltage by the value of the resistance (Z = R), we get the instantaneous value of the electric current (Eq. 4.3). If we replace u with the sine function, we get:

$$i = \frac{u}{R} = \frac{U_0 \cdot sen\, \omega t}{R} = \frac{U_0}{R} \cdot sen\, \omega t(A) \tag{4.3}$$

Obviously, if the instantaneous voltage value equals the maximum value U_0, the result of dividing it by the resistance is the current I_0:

$$I_0 = \frac{U_0}{R} \tag{4.4}$$

As a result, we express the instantaneous value of the current i, as:

$$i = I_0 \cdot sen(\omega t) \tag{4.5}$$

We can plot the phasor diagram of voltage (U) and current (I) (Fig. 4.4). We graph these quantities as two vectors rotating at speed w, counterclockwise. The Y-axis projection of voltage and current waves corresponds to the instantaneous values of current and voltage.

As a result, it is possible to conclude that the current oscillates in tandem with the applied voltage. Specifically, the voltage and current fluctuate simultaneously (although with distinct amplitudes), resulting in a phase relationship between them (Fig. 4.5).

For practical calculations, it is crucial to use the effective values of voltage and current, known as U and I. The rms values are proportional to the maximum values of current and voltage by $\sqrt{2}$, and we express it as

$$U_0 = \sqrt{2} \cdot U(V) \tag{4.6}$$

$$I_0 = \sqrt{2} \cdot I(A) \tag{4.7}$$

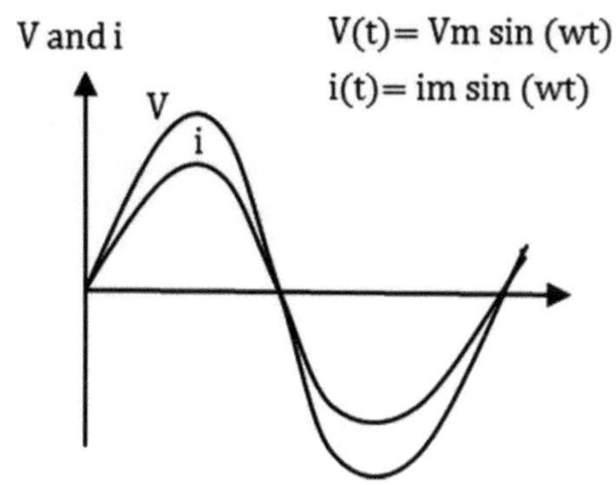

Fig. 4.4 Voltage and current waves in a resistive circuit. *Source* Own elaboration

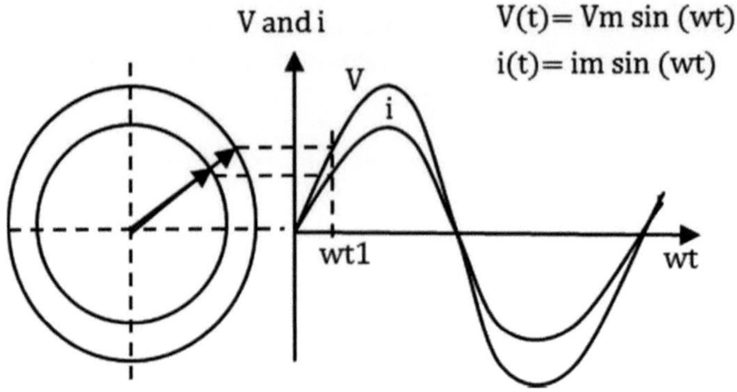

Fig. 4.5 Voltage and current Phasors. *Source* Own elaboration

Substituting Eqs. 4.6 and 4.7 into 4.2 yields:

$$\sqrt{2} \cdot I = \frac{\sqrt{2} \cdot U}{R} \tag{4.8}$$

This brings us to Ohm's law:

$$I = \frac{U}{R} \tag{4.9}$$

When we have a circuit where a current with a rms value, I passes through a pure ohmic resistor, a voltage drop occurs. Then, we know it as the "resistance drop or active voltage drop", which is expressed as follows by Ohm's law (Eq. 4.10):

$$U_R = I \cdot R \tag{4.10}$$

4.3 Purely Capacitive Circuit

Now we examine the charging and discharging process of a capacitor with AC sources. When we apply a sinusoidal alternating voltage to a fully discharged capacitor, a strong charging current appears during the first quarter of cycle. Once the current source reverses the polarity, the capacitor discharges the stored energy, and so on (Fig. 4.6).

Question What happens to the voltage and current once the capacitor has been charged?

Once the capacitor has been totally charged, we know that the current will reach zero and the voltage will be at its maximum.

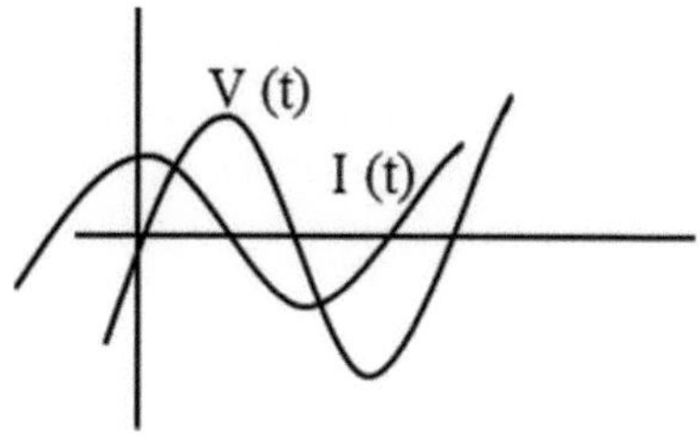

Fig. 4.6 Voltage and Current through a capacitor. *Source* Own elaboration

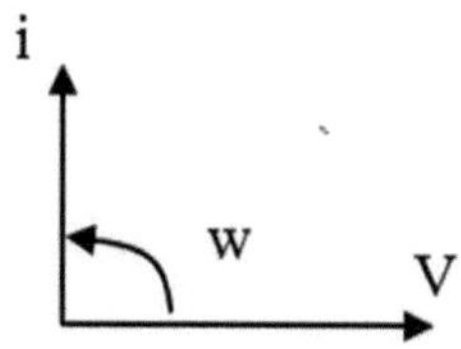

Fig. 4.7 Voltage and current phasor diagram. *Source* Own elaboration

When we plot the phasor diagram, the current waveform is 90 degrees ahead of the voltage waveform (Fig. 4.7).

Expressing the sinusoidal voltage as

$$V(t) = V_m sin(wt)(V) \tag{4.11}$$

Then, the voltage in the capacitor is also

$$V_c(t) = V_m sin(wt)(V) \tag{4.12}$$

According to the constitutive law of the capacitor

$$I_c(t) = I_m sin(wt + 90^\circ)(A) \tag{4.13}$$

This means that the capacitor advances the current by 90º with respect the voltage. Once the voltage source that feeds the capacitor falls, it discharges the stored energy during the previous quarter cycle, causing a discharge current to appear in the circuit.

Now on we will answer questions like:

Question What is the charge-discharge time of a capacitor?, and what is the average electricity consumption?

Because of the behavior of the capacitor, when we plot the current on the vector diagram (Fig. 4.7), we draw it at a 90° angle to the voltage. The average power that the capacitor absorbs is zero because it charges within one quarter of the cycle and returns it in the next quarter of the cycle. A pure capacitor advances the current by an angle of 90° to the voltage.

Question Do you think the alternating current "I", will pass through the capacitor's dielectric?

The alternating current that flows through a capacitor generates a sinusoidal electric current in the circuit. The continuous and alternating charging and discharging of the capacitor generates a sinusoidal electric current that flows through the circuit. We have to keep in mind that this current does not pass through the dielectric of the capacitor.

In order to determine the value of the current, we will use a capacitor with a capacity C that connects to the terminals of an alternator that delivers sinusoidal alternating voltage. When we designate the capacitor charge at instant t as q, we obtain:

$$q = c \cdot U \tag{4.14}$$

And as we know that:

$$i = \frac{dq}{dt} \tag{4.15}$$

Combining Eqs. 4.14 and 4.15 yields

$$i = \frac{d(C \cdot u)}{dt} = C \cdot \frac{du}{dt} = C \cdot \frac{d(U_0 \cdot sen\omega t)}{dt} \tag{4.16}$$

And therefore:

$$i = C \cdot \omega \cdot U_0 \cdot cos(\omega t) \tag{4.17}$$

However, as we know:

$$cos(\omega t) = sen\left(\omega t + \frac{\pi}{2}\right) \tag{4.18}$$

It can be defined as:

$$i = C \cdot \omega \cdot U_0 \cdot sen\left(\omega t + \frac{\pi}{2}\right) \tag{4.19}$$

Remember that, according to the general expression of a sine alternating current wave, we obtain the instantaneous value of the current "I" as the projection on the y-axis of a rotating vector with a modulus equal to the maximum value, I_0, which rotates counterclockwise.

Therefore, it follows that:

$$I_0 = C \cdot \omega \cdot U_0 \tag{4.20}$$

And thus, we can express the immediate value "I":

$$i = I_0 \cdot sen\left(\omega t + \frac{\pi}{2}\right) \tag{4.21}$$

This equation shows that the current in a circuit with pure capacitance is a sine function of time, with the same frequency as the voltage source, but leading it by π/2 radians.

Note that we can express Eq. 4.21 as a function of voltage:

$$I_0 = C \cdot \omega \cdot U_0 = \frac{U_0}{\frac{1}{C \cdot \omega}}, \tag{4.22}$$

Then, we have

$$X_C = \frac{1}{C \cdot \omega}$$

We can plot the time-domain sinusoidal waveform as well as a rotating phasor diagram (Fig. 4.8).

We find the maximum current I_0 using Ohm's Law by dividing the maximum voltage value U_0 by the opposition to current passage when the instantaneous voltage value u equals U_0. We can place a capacitor that connects to an alternating current voltage source (Fig. 4.9).

Therefore, capacitive reactance or capacitance, denoted as Xc, characterizes the opposition presented to the flow of an alternating current by a capacitance of a circuit.

Taking the real part, we have:

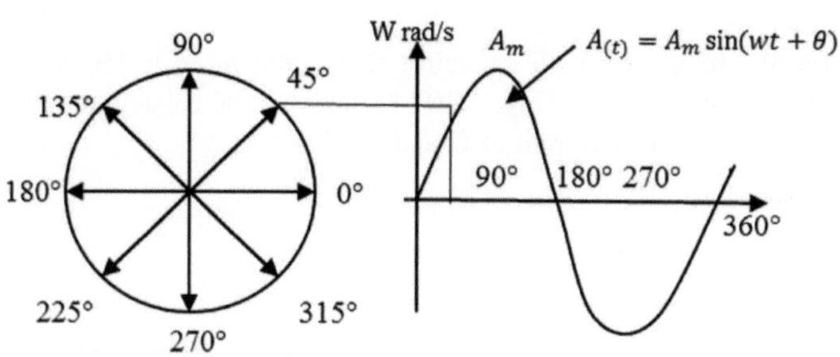

Fig. 4.8 Sinusoidal waveform in the time domain, with rotating phasor. *Source* Own elaboration

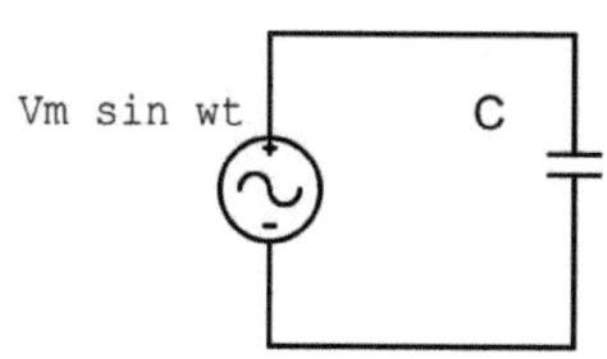

Fig. 4.9 Capacitor connected to a sinusoidal voltage source. *Source* Own elaboration

$$I = \frac{U}{X_C} \tag{4.23}$$

In this case, the circuit impedance equals the value of the capacitive reactance (Eq. 4.1), using Ohm's Law for AC ($Z = X_C$). We see that higher capacitor capacitance C results in lower capacitive reactance (Eq. 4.22). We also see that lower X_C results in higher capacitor charging and discharging currents (Eq. 4.23). The higher the angular frequency of the sinusoidal current "w", the faster the capacitor charges and discharges and thus the current through the circuit increases.

It is important to keep in mind that the w is directly proportional to the frequency. As the frequency increases, X_C will decrease. In an alternating current circuit, a capacitor represents a small resistance or a short circuit for high-frequency currents and a barrier for low-frequency currents. We could define direct current as having zero-frequency. It is to highlight that.

We measure regular frequency (f) in cycles per second (cps) or hertz (Hz), and angular frequency (w) in radians per second.

Then we know how to convert:

$$w = 2\pi f$$

The difference of voltage we have between the plates of a capacitor, that appears after current I, is known as the "capacitive voltage drop" or "capacitive reactive voltage". We calculate this quantity as

$$U_c = X_c \cdot I = \frac{I}{C \cdot \omega} \tag{4.24}$$

Problem 4.1 We have a 175 μF capacitor that connects to a 230 V AC voltage source of a frequency of 50 Hz (Fig. 4.10). Determine the capacitive reactance and the current. Draw a vector diagram of U and I.

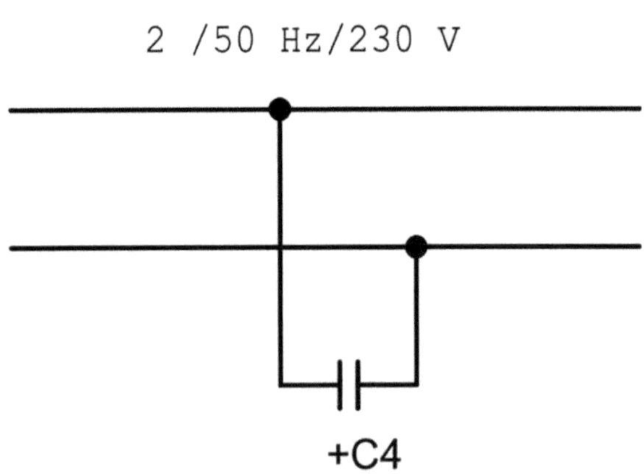

Fig. 4.10 Problem 4.1 Circuit with only a capacitor. *Source* Own elaboration

4.4 Purely Inductive Circuit

Let us connect the terminals of a sinusoidal alternating current generator and an ideal inductance. Similarly to capacitors, we can draw an alternating current source that connects to a coil (Fig. 4.11).

The voltage at the generator or inductance terminals is the same since there are not losses due to voltage drops.

Its mathematical expression is:

$$u = U_0 \cdot sen(\omega t) \tag{4.25}$$

In any circuit where varying current flows, it appears an induced emf. The value of this emf source, which we express it in volts, is the product of the self-induction coefficient L, given in Hertz, and the rate of change of the current over time $\frac{di}{dt}$, given in amperes. Recalling the expression of the induced emf, e_L:

$$e_L = -L \cdot \frac{di}{dt} \tag{4.26}$$

The minus sign in Lenz's law shows that the induced emf, e_L, in any coil because of a changing magnetic flux or field strength opposes the change that produces it.

Applying Kirchhoff's second law to the circuit, we get:

$$u + e_L = 0 \tag{4.27}$$

And substituting Eqs. 4.26 into (4.27) we have

$$u + \left(-L \cdot \frac{di}{dt}\right) = 0 \tag{4.28}$$

$$u = L \cdot \frac{di}{dt} \tag{4.29}$$

Substituting the instantaneous voltage value:

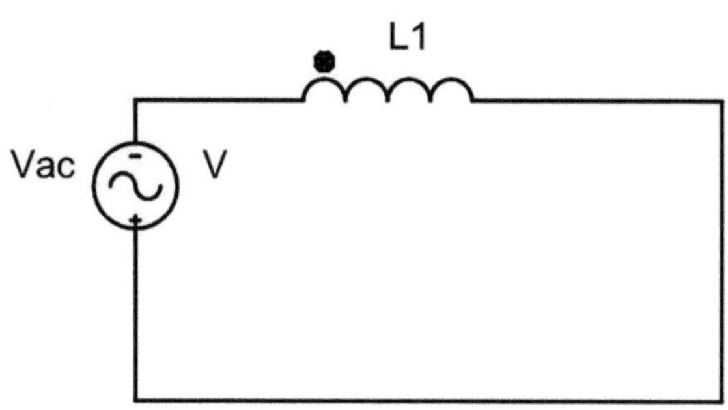

Fig. 4.11 Purely inductive circuits. *Source* Own elaboration

$$U_0 \cdot sen(\omega t) = L \cdot \frac{di}{dt} \tag{4.30}$$

Separating the variables in the differential equation yields:

$$di = \frac{U_0}{L} \cdot sen(\omega t) \cdot dt \tag{4.31}$$

Only by integrating it we obtain:

$$\int di = \frac{U_0}{L} \cdot \int sen(\omega t) \cdot dt \tag{4.32}$$

Therefore, it results in:

$$i = -\frac{U_0}{L\omega} \cdot cos(\omega t) = \frac{U_0}{L\omega} \cdot sen\left(\omega t - \frac{\pi}{2}\right) \tag{4.33}$$

Since:

$$\pm cos(\omega t) = sen\left(\omega t \pm \frac{\pi}{2}\right) \tag{4.34}$$

Now:

$$I_0 = \frac{U_0}{L\omega} \tag{4.35}$$

Finally, we have:

$$i = I_0 \cdot sen\left(\omega t - \frac{\pi}{2}\right) \tag{4.36}$$

Now we plot the voltage and current waveform of a purely inductive circuit (Fig. 4.12).

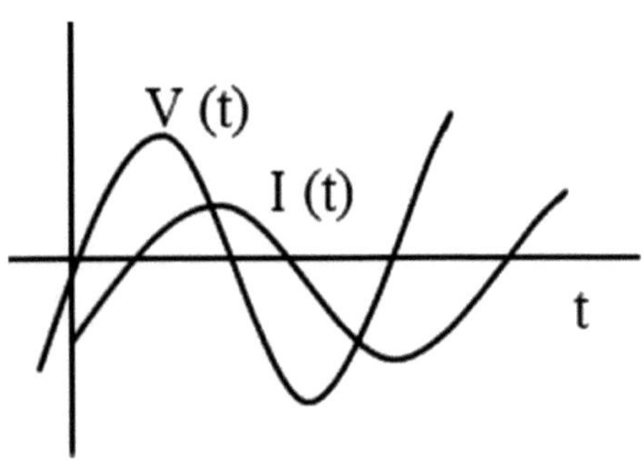

Fig. 4.12 Voltage and current through an inductance. *Source* Own elaboration

The current is then a sinusoidal function having the same frequency as the applied emf, but offset by π/2 radians from that voltage. Substituting (Eqs. 4.6 and 4.7) into Eq. 4.35 yields:

$$\sqrt{2} \cdot I = \frac{\sqrt{2} \cdot U}{\omega \cdot L} \tag{4.37}$$

Therefore, its root-mean-square (rms) value will be:

$$I = \frac{U}{\omega \cdot L} \tag{4.38}$$

Substituting ωL in Eqs. 4.35 and 4.38 yields:

$$\omega \cdot L = \frac{U_0}{I_0} = \frac{U}{I} = X_L(\Omega) \tag{4.39}$$

When we express L in Henry and f in Hertz, and the units of $\omega \cdot L = ohms$. X_L, also known as inductive or self-induction reactance, serves the same purpose as resistance and capacitive reactance in previous circuits by representing the coil's opposition to the passage of current.

Question Does the Joule effect occur?

The Joule effect does not occur under a reactive load. It varies with the frequency, f, and also creates a –π/2 radian phase difference between voltage and current. Now we plot the voltage, magnetic flux and current waveforms in a purely inductive circuit (Fig. 4.13).

One thing to keep in mind is that the current does not immediately follow the variations of the voltage, but rather exhibits "inertia" and lags behind the applied voltage. Inductances resist the flow of alternating currents through a reactance X_L, which is proportional to the frequency (remember that ω = 2πf). Recalling Ohm's law for AC, in this case, the impedance of the circuit corresponds with the value of

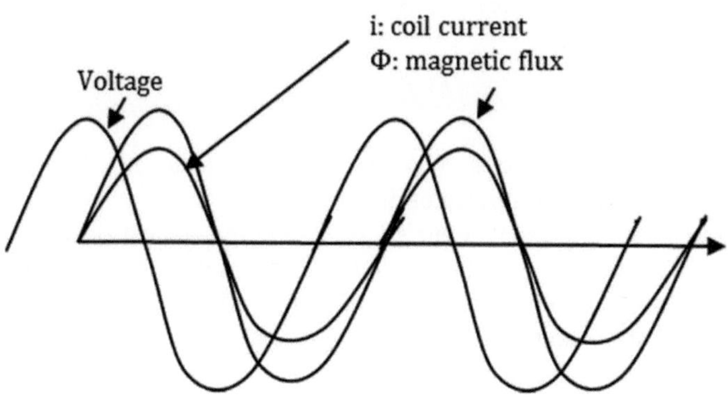

Fig. 4.13 Voltage, magnetic flux and current waveform through a coil. *Source* Own elaboration

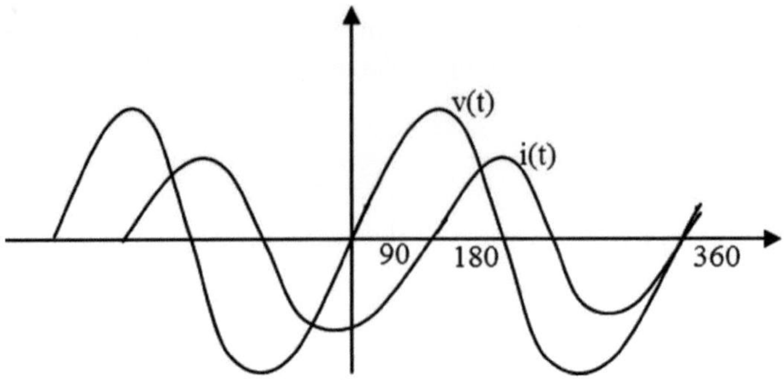

Fig. 4.14 Behavior of the current wave in inductive circuit. *Source* Own elaboration

the inductive load Z = X_L. We call "inductive voltage drop or inductive reactive voltage" to the voltage drop across the coil when the current I flow through it and we express it as:

$$U_L = IX_L \tag{4.40}$$

When we connect a sinusoidal AC voltage source to an inductance, we can describe its behavior. Then, current wave lag behind the voltage wave in an inductive circuit (Fig. 4.14).

When a sinusoidal AC current flows through the coil, the coil's magnetic flux lines intersect its own conductors.

4.5 Faraday's Law

Faraday discovered that when a wire coil enters a magnetic field, it generates an emf. The induced current generates a magnetic field that counteracts the magnetic field buildup in the coil. This is expressed as:

$$Emf = N\frac{\Delta\Phi}{\Delta t} \tag{4.41}$$

where N is the number of loops, Φ = BA is the magnetic flux, B is the external magnetic field, and A is the surface of the loop.

The negative sign represents Lenz's Law within the context of electromagnetic induction. As well reviewed, emf refers to the electromotive force that appears. The current through the circuit generates a flux and flux linkage, resulting in an emf, according to Faraday's and Lenz's Laws (Fig. 4.15).

This creates a self-induced emf that, according to Lenz's law, opposes the action that caused it. It resists changes in the current under all circumstances. According to Lenz's law, we can draw the relationship between current and induction in a coil (Fig. 4.16).

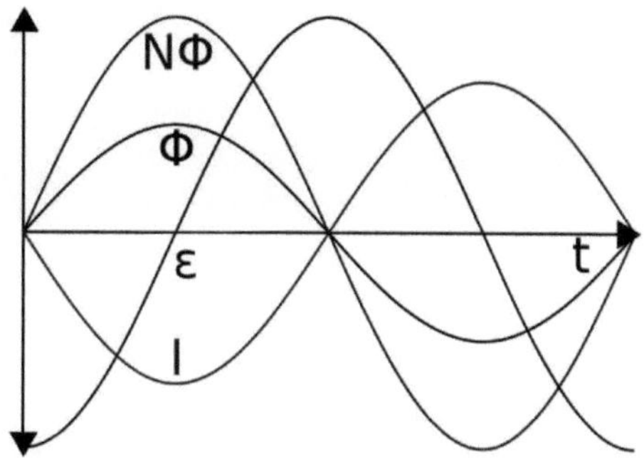

Fig. 4.15 Faraday's and Lenz's Law. *Source* From wikicommons under free to use license [1]

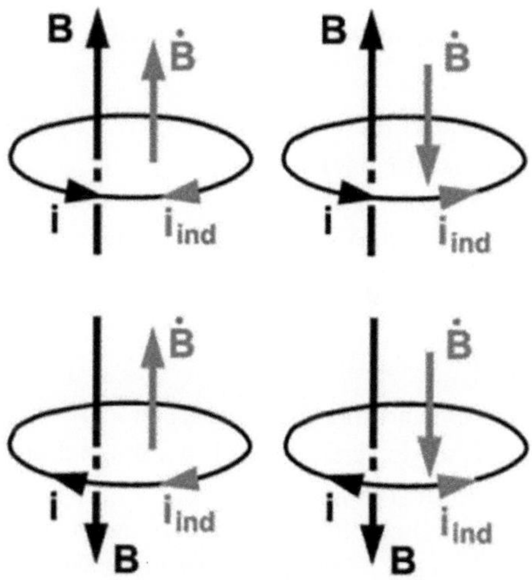

Fig. 4.16 Induced current opposes primary current. Original. *Source* From wikicommons under CC 2.0 use license [2]

The magnetic field increases as the current increases. As a result, an opposing emf appears. This causes the electric current to lag behind the voltage. In this scenario, the coil absorbs energy in the form of an increasing magnetic field. Notice that, once the current has stabilized at its peak level through the coil, the self-induction emf is zero. However, as the current decreases, so does the field, resulting in a self-induction emf that resists the decreasing current.

Question What is the energy behavior of the coil during this phase?

Now, the coil releases the energy it had stored in the previous quarter cycle to the generator as a diminishing magnetic field. This behavior differs from that of the capacitor, despite their similarities. Now, we plot the voltage and current waves travel in an inductive circuit (Fig. 4.17).

We see in this vector diagram (Fig. 4.17) the electric current with a 90° delay with respect to the voltage. It is clear that as the current rise, the voltage reaches a peak, charging the coil.

What is the average power consumption value?

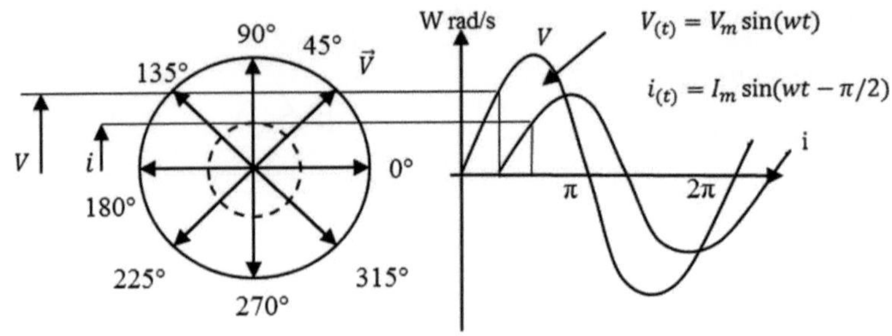

Fig. 4.17 Graph current opposes primary current. *Source* Own elaboration

4.6 Sinusoidal RL Circuit

We restrict the current flow when we connect a resistor and an inductive reactance and X-L (Fig. 4.18).

As the two elements are series-connected, the current that flows is similar for both elements. Ohm's law applies in this case, which we express as:

$$I = \frac{U}{Z}$$

We determine the impedance Z by combining the impact of R and X_L. Since R and L are in series, there is a voltage drop in both elements. The total voltage is the sum of the two.

Applying Ohm's law the voltage drops are:

$U_R = R \cdot I$ (Active or resistance voltage drop)

$U_L = X_L \cdot I$ (Inductive voltage drop or inductive reactive voltage)

And the sum must equal the total voltage U.

We will use vector diagrams to establish the equivalence between these quantities. Then, the total phasor voltage U is the sum of the two voltage phasors, and we express

Fig. 4.18 Sinusoidal R-L circuit. *Source* Own elaboration

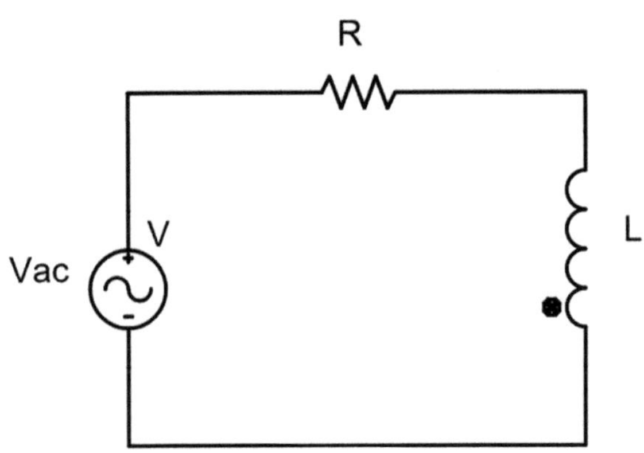

it as

$$\vec{U} = \vec{U_R} + \vec{U_L} \tag{4.42}$$

We follow the next steps to construct the vector diagram:

(1) We build the vector diagram with the reference current at the position of wt = 0, as it is the usual case for R and L.
(2) We place U_R on the vector diagram parallel to the current because resistance causes no phase shift.

Question How do we draw the vector diagram?

(3) The voltage UL is 90° ahead the current (delaying the current with respect to the voltage means advancing the voltage with respect to the current). U_R and U_L are the legs of a right angle, and U the hypotenuse of it. Then we can write:

$$U = \sqrt{U_R^2 + U_L^2} \tag{4.43}$$

The voltage triangle also yields the angle phi, which represents the phase difference between the voltage at the generator terminals (U) and the current flowing through the circuit (I). Then we have:

$$\varphi = arctg \frac{U_L}{U_R} \tag{4.44}$$

We obtain the impedance triangle, which is similar to the voltage triangle by dividing its side by I. In this triangle the following holds true:

$$Z = \sqrt{R^2 + X_L^2} \tag{4.45}$$

$$\varphi = arctg \frac{X_L}{R} \tag{4.46}$$

4.7 Sinusoidal RC Series Circuit

When a resistor and a capacitor connect in series (Fig. 4.19), the combined effect of R and X_C limits the current, as in the RL circuit. We can use Ohm's law to determine its value.

This law finds the impedance Z by combining the effects of R and X_C. Then, we have:

$$I = \frac{U}{Z}$$

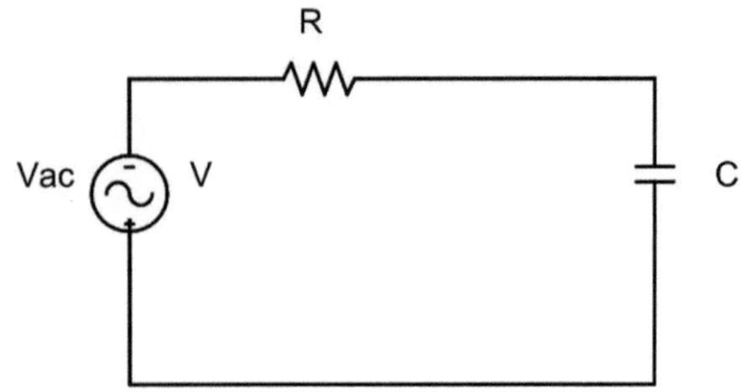

Fig. 4.19 Sinusoidal R-C circuit. *Source* Own elaboration

What is the difference in Z and angle phi with an RL circuit? To find the value of Z and phi, we follow the same steps as in the previous case. However, the impedance now comprises the composition of the vectors R and X_C, and the resulting phase shift will be a voltage delay.

Because R and C are in series, then there will be a voltage drop in each component. Therefore, the sum of the two voltage drops equals the total voltage. The voltage across each component is

$$V_R = R \cdot I \quad \text{(Active or Resistance Voltage Drop)} \tag{4.47}$$

$$V_C = X_C \cdot I \quad \text{(Capacitive voltage drop or capacitive reactive voltage)} \tag{4.48}$$

And their sum must equal the total voltage V. We use vector diagrams to show the equivalence between these quantities. The voltage V is the sum vector of these two voltages which can be plotted in the phasor diagram of a series RC circuit (Fig. 4.20).

$$\vec{V} = \vec{V_R} + \vec{V_C} \tag{4.49}$$

To create the vector diagram, follow these steps. Begin by drawing the vector diagram with the current as a reference point in position wt = 0, as this is typical for both R and C.

Question What steps must we follow to construct the vector diagram?

1. We place V_R on the vector diagram aligned with the current because the resistance does not cause any phase shift.

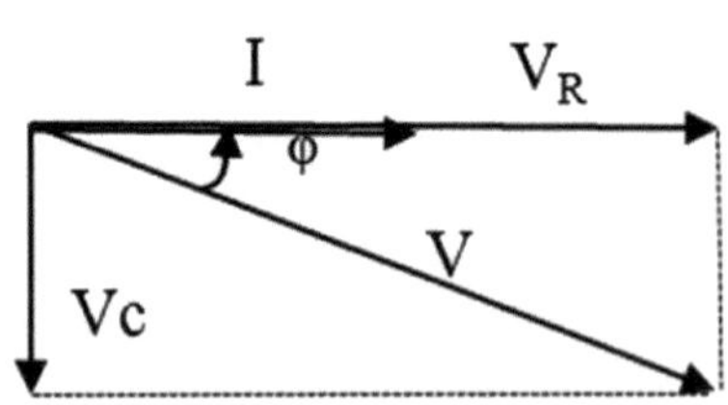

Fig. 4.20 Phasor diagram of a series R-C circuit. *Source* Own elaboration

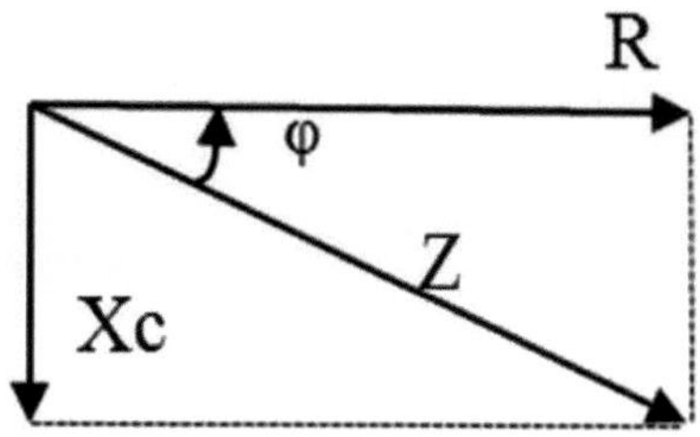

Fig. 4.21 Impedance triangle. *Source* Own elaboration

2. The voltage V_C is 90º out of phase with the current (advancing the current in phase with the voltage means delaying the voltage in phase with the current).

The vector addition of V_R and V_C, gives the resultant voltage V, which is phase-shifted by an angle fi relative to the current.

V_R and V_C compose the legs of a right angle, and V is the hypotenuse of it. It means that:

$$V = \sqrt{V_R^2 + V_C^2} \tag{4.50}$$

The voltage triangle also yields the phase difference angle between the generator terminal voltages V and the current through the circuit I. This angle is calculated as:

$$\varphi = arctg\frac{V_C}{V_R} \tag{4.51}$$

Similarly, the impedance triangle of the circuit, which includes R, and X_C^2 can be drawn (Fig. 4.21).

The impedance triangle, which is similar to the voltage triangle, can be obtained by dividing the values of its sides by I. The following holds true in the RC circuit impedance triangle:

$$Z = \sqrt{R^2 + X_C^2} \tag{4.52}$$

$$\varphi = arctg\frac{X_C}{R} \tag{4.53}$$

4.8 RLC Series Circuit

When a resistor, a coil and a capacitor connect in series, the combined effect of R, X_L and X_C limits the current. We use Ohm's law to determine the value of the current. Combining the effects of R, X_L, and X_C determines the impedance Z. We use the same previous procedure to determine the values of Z and fi. The overall impedance

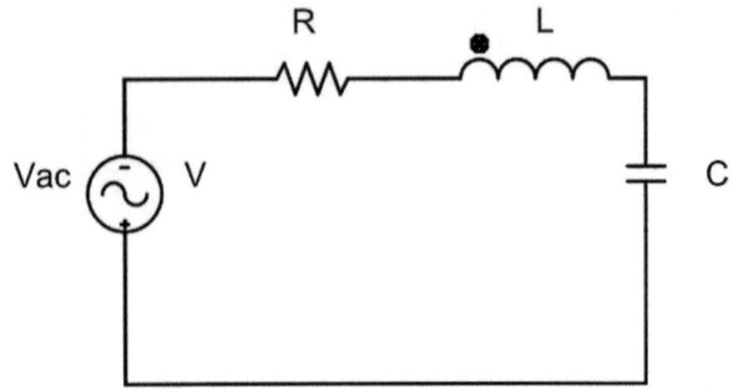

Fig. 4.22 R-L-C series circuit. *Source* Own elaboration

of the circuit is R, X_L, and X_C. Note that because the three elements are in series, each will experience a voltage drop. As a result, the sum of the three voltage drops will result in the total voltage. A R-L-C series circuit that includes a resistor, an inductance and a capacitor (Fig. 4.22).

The voltage drops in each component are:

$U_R = R \cdot I$; This is the active or resistance voltage drop.

$U_L = X_L \cdot I$; This is the inductive voltage drop or inductive voltage.

$$U_C = X_C \cdot I$$

This is the capacitive voltage drop or capacitive reactive voltage. And their sum must equal the total voltage U.

Question What steps must we take to draw the vector diagram?

To establish the equivalence between these quantities, we will use vector diagrams again. According to Kirchhoff voltage law, U becomes the vector addition of these three voltages (Fig. 4.23):

$$\vec{U} = \vec{U_R} + \vec{U_L} + \vec{U_C} \tag{4.54}$$

To draw the vector diagram, we follow the following steps:

1. We draw the vector diagram with the current as a reference at wt=0 because it is common to R, L and C.
2. We draw U_R on the vector diagram in phase with the current because the ohmic resistance does not cause a phase shift.
3. The voltage U_L is 90° ahead of the current.
4. The voltage U_C delays 90° regarding to the current.

It is noted that the vector diagram shows an inverse relationship between the capacitor and coil voltage drops. The vector sum of U_R, U_L, and U_C determines the system voltage U, which advances by an angle fi with respect to the current.

We can derive the voltage triangle from the vector diagram of the R-L-C circuit.

Because U_R, and (U_L – U_C) compose the legs of a right triangle and U the hypotenuse of the triangle, it satisfies:

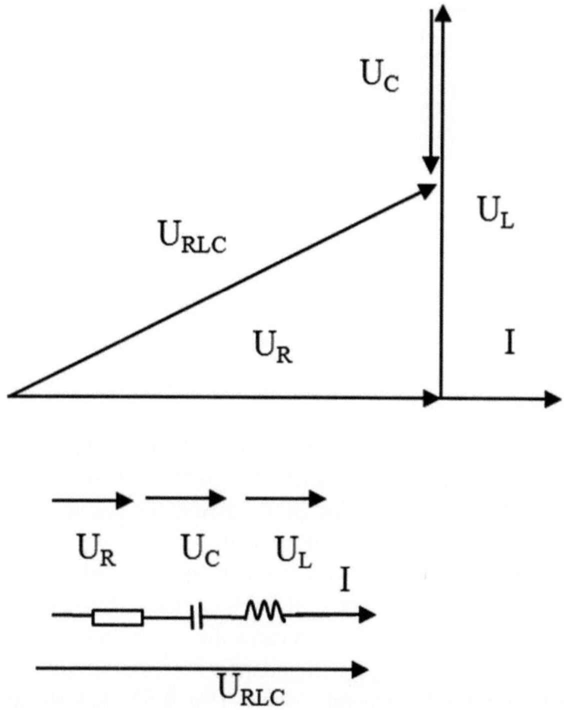

Fig. 4.23 Phasor diagram of a series RLC circuit. *Source* Own elaboration

$$U = \sqrt{U_R^2 + (U_L - U_C)^2} \tag{4.55}$$

The voltage triangle determines the angle phi, which corresponds to the phase difference between the voltage at the generator's terminal, U, and the current that flows through the circuit, I. The angle can be easily found (Eq. 4.56).

$$\varphi = arctg\frac{(U_L - U_C)}{U_R} \tag{4.56}$$

If we divide the sides of the voltage triangle by I, the resulting impedance triangle is similar. Looking at the impedance triangle we have:

$$Z = \sqrt{R^2 + (X_L - X_C)^2} \tag{4.57}$$

and

$$\alpha = arctg\frac{X_L - X_C}{U_R} \tag{4.58}$$

Then we can draw the impedance triangle of the R-L-C series circuit (Fig. 4.24).

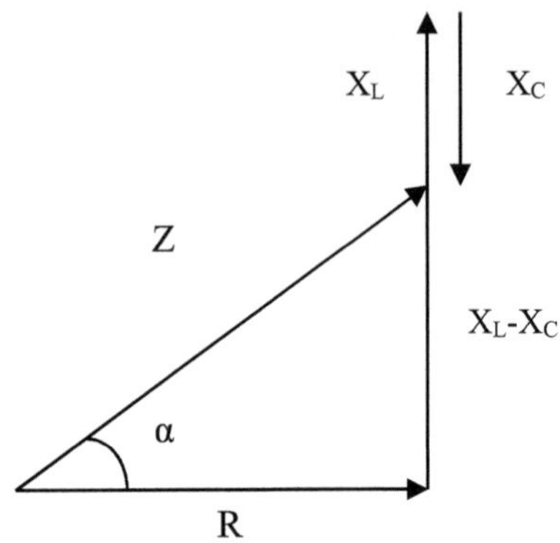

Fig. 4.24 Impedance triangle of the RLC series circuit. *Source* Own elaboration

In the impedance triangle we see that the capacitor partially compensates for the coil's reactance. This causes the circuit to behave like an inductive circuit, and this provokes a current delay (fi) relative to the voltage. Then, the capacitor compensates for some of the effects of the coil.

Question How do a coil and a capacitor behave in an RLC circuit?

In this scenario, the coil discharges the stored energy, some of which charges the capacitor. The capacitor returns the energy to the coil during the following quarter cycle. We express the impedance Z of the circuit as a function of R, L, and C by substituting the resistors' values (Eq. 4.57). Recalling that:

$$X_c = \frac{1}{C \cdot \omega} \tag{4.59}$$

and:

$$X_L = \omega \cdot L \tag{4.60}$$

We calculate the equivalent impedance (Eq. 4.61):

$$Z = \sqrt{R^2 + (X_L - X_C)^2} \tag{4.61}$$

Substituting Eqs. 4.59, and 4.60 we have (Eq. 4.62):

$$Z = \sqrt{R^2 + \left(\omega \cdot L - \frac{1}{C \cdot \omega}\right)^2} \tag{4.62}$$

By operating similarly, we represent the angle of this impedance as:

$$\varphi = arctg \frac{\omega \cdot L - \frac{1}{\omega \cdot C}}{R} \tag{4.63}$$

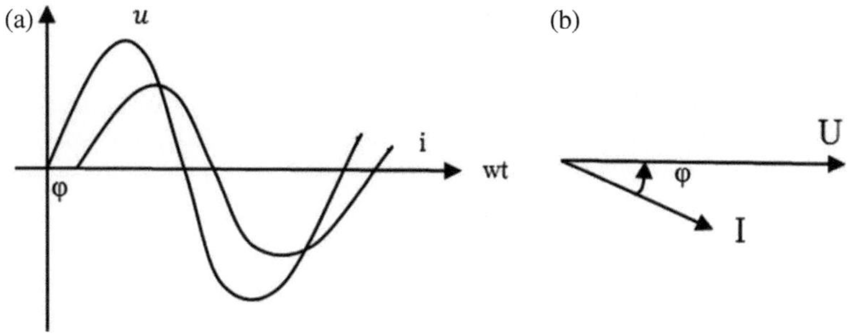

Fig. 4.25. **a**, and **b** Cartesian and phasor diagram of a R-L-C series circuit. *Source* Own elaboration

In the steady state, the instantaneous value of the current will be out of phase with the applied voltage. The difference of angle phase, phi, results in the circuit's impedance. The voltage-impedance triangle shows this angle. Then we can determine the instantaneous value (Eq. 4.64):

$$i = I_0 \cdot sen(\omega \cdot t - \varphi) \tag{4.64}$$

Whose maximum value is:

$$I_0 = \frac{U_0}{\sqrt{R^2 + \left(\omega \cdot L - \frac{1}{C \cdot \omega}\right)^2}} \tag{4.65}$$

The current that appears is an alternating current having the same frequency as the voltage source. It also undergoes a phase shift of angle φ (Eq. 4.63). According to this Equation, we plot the current and voltage wave (Fig. 4.25a), as well the phasor diagram of voltage and current in this circuit (Fig. 4.25b).

4.9 Summary

This chapter analyzes the behavior of fundamental passive components—resistors (R), inductors (L), and capacitors (C)—in alternating current (AC) circuits. It begins by establishing Ohm's law for sinusoidal steady-state conditions. The chapter then examines each component in isolation, detailing the characteristics of purely resistive, purely capacitive, and purely inductive circuits.

The core of the chapter explores how these elements behave in combination. It systematically analyzes the response of RL, RC, and RLC series circuits to sinusoidal excitation, explaining how the unique properties of inductors and capacitors (including concepts like Faraday's Law) affect the overall circuit's voltage, current, and phase relationship.

References

1. Sjlegg, English: Graph of current, which induces a flux and flux linkage, which induces an emf, to illustrate Faraday's and Lenz's Laws (2009). Accessed 10 Dec 2023. [Online]. https://commons.wikimedia.org/wiki/File:Induction_graph.svg
2. P. wormer, English: Lenz' law of electromagnetism (2010). Accessed 10 Dec 2023. [Online]. https://commons.wikimedia.org/wiki/File:Lenz_law.png

Chapter 5
Alternating Current Power and Measurement

Abstract This chapter explores the concept of power. Specifically, it focuses in the analysis of power in alternating current circuits. From the analysis of power in purely resistive circuits to the analysis of power in purely inductive circuits, this chapter analyses the power triangle of R-L circuit, and draws the vector diagram of RL circuits. This chapter finishes by analyzing the instantaneous power in capacitor and inductances. We will review the reactive, active, and apparent power, and the power factor.

Keywords Power in ac circuits · Power in resistive circuits · Power in capacitive circuits · Power in inductive circuits · Power factor

5.1 Introduction. What Is Power?

In Chap. 3, we defined power as the measure of the amount of effective work completed. We measure DC power in watts. In this chapter, we will answer to questions like how much energy does the coils and capacitors consume from the grid.

The coils and capacitors draw energy for only a quarter of a cycle before releasing it in the next cycle, resulting in an average zero power consumption. In AC we also measure power in watts. But,

Question What is the value of one watt?

We know that the battery delivers power when the current I flows out of the positive voltage terminal of the battery. The power then is positive, and we have

$$W_{AB} = U_{AB} \times q \tag{5.1}$$

$$P = \frac{W}{t} = \frac{Uq}{t} = UI \tag{5.2}$$

Then, we see that one watt is equal to the product of one ampere and one volt.

D. Borge-Diez et al., *Fundamentals of Electrical Power Engineering*, Lecture Notes in Electrical Engineering 1540, https://doi.org/10.1007/978-3-032-18904-2_5

5.2 Power in a Purely Resistive Circuit

In AC circuits, we use the instantaneous values of voltage and current to find the instantaneous power. To better understand this concept, we plot curves based on the instantaneous power of receivers. As in DC circuits, we start by determining the current in the circuit. We will analyze the power in the resistive circuit (Fig. 5.1).

The value of instant current is:

$$i = I_0 \cdot sen(\omega t) \tag{5.3}$$

In addition, the instantaneous voltage function has the same frequency:

$$u = U_0 \cdot sen(\omega t) \tag{5.4}$$

We can also express these equations as:

$$i = \sqrt{2} \cdot I \cdot sen(\omega t) \tag{5.5}$$

$$u = \sqrt{2} \cdot U \cdot sen(\omega t) \tag{5.6}$$

In order to find the power, we have to multiply voltage and current. Now we plot the voltage wave as well the power waveform in a resistive load (Fig. 5.2). When the voltage is zero, the power is also zero. Otherwise, the power waveform is always positive, because the resistor can only dissipate energy, not supply energy to the circuit. The resistor outputs power regardless of whether the voltage is positive or negative.

We get the instantaneous power curve multiplying the sine function of current $\left(i = \sqrt{2} \cdot I \cdot sen\omega t\right)$ by the sine function of voltage $\left(u = \sqrt{2} \cdot V \cdot sen\omega t\right)$. As we see, this curve is always positive.

Question What does this curve show about resistance?

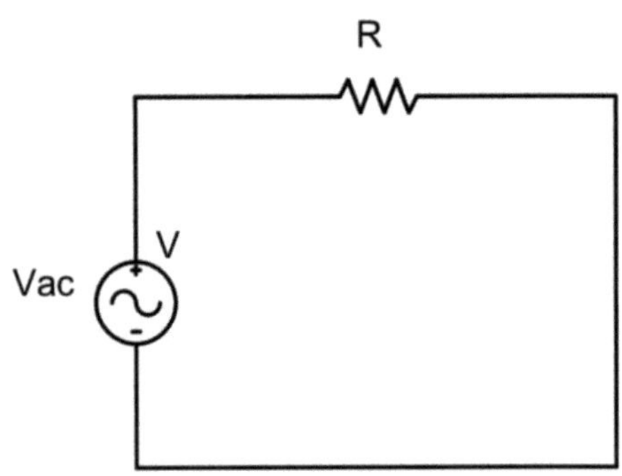

Fig. 5.1 Purely resistive circuit with AC voltage source. *Source* own elaboration

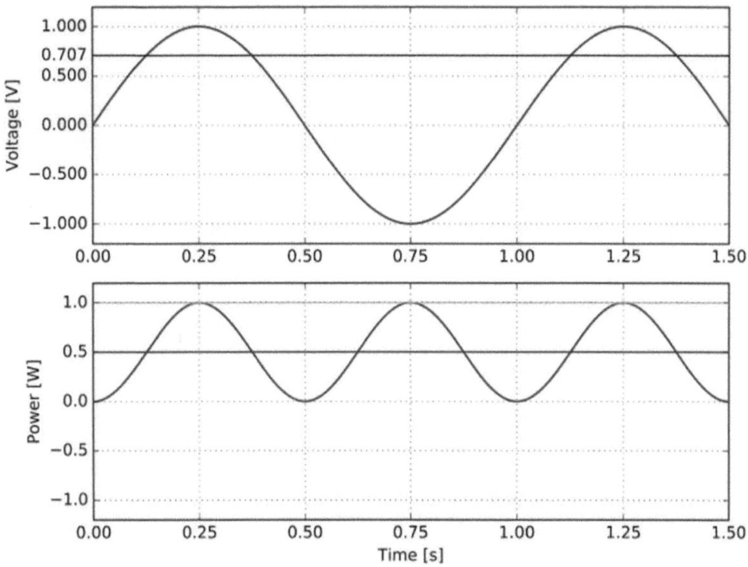

Fig. 5.2 AC power to its real. Original. *Source* from wikicommons under CC 2.0 use license [1]

This graph illustrates that the resistor absorbs energy from the generator and converts it into heat.

Question What does the average value of this curve represent?

The mean value of this power is the active power (P) or real power, and we measure it with a wattmeter.

In a resistor we know that power is:

$$P = R \cdot I^2 (W) \tag{5.7}$$

In this case, the instantaneous power equals the product of instantaneous current and voltage. Multiplying, we get

$$\begin{aligned} p(t) &= u(t) \cdot i(t) = \sqrt{2} \cdot V \cdot sen\omega t \cdot \sqrt{2} \cdot I \cdot sen\omega t = 2 \cdot U \cdot I \cdot sen^2 \omega t \\ &= U \cdot I \cdot (1 - \cos 2\omega t) \end{aligned} \tag{5.8}$$

Applying trigonometric identities, we know:

$$sen^2 u = \frac{1 - cos2u}{2}$$

$$P = \frac{1}{T} \int_0^T p \cdot dt = U \cdot I \tag{5.9}$$

Then, we obtain the average power integrating the power over one cycle and dividing by the cycle time.

$$\begin{aligned} P &= \frac{1}{T}\int_0^T p \cdot dt = U \cdot I \cdot (1 - cos2wt) \cdot dt = \frac{UI}{T}\int_0^T dt \\ &- U \cdot I \cdot \int_0^T cos2wtdt = \frac{UI}{T}[t]_0^T - \frac{UI}{2w}[sen2wt]_0^T \\ &= \frac{U \cdot I}{T} \cdot T = U \cdot I \end{aligned} \tag{5.10}$$

$$\text{Remember} : \int \cos 2xdx = \frac{1}{2}\sin 2x \text{ (integrating parts)} \tag{5.11}$$

$$\begin{aligned} \int \cos 2xdx &= \frac{1}{2}\int \cos udu \\ \int \cos udu &= \sin u \\ u = 2x \rightarrow \frac{du}{dx} &= 2 \rightarrow dx = \frac{1}{2}du \\ &= 2x \rightarrow \frac{du}{dx} = 2 \\ \int \cos 2xdx &= \frac{1}{2}\sin u \end{aligned}$$

Therefore, the most significant measure is the active power, which comes from the effective values of U and I.

5.3 Power in Circuits with Coil

Now we analyze the power in circuits with a single coil. We know that current delays to the voltage by 90 degrees, or $\pi/2$ radians (Fig. 5.3). Consequently, the instantaneous values of current and voltage are:

$$u = \sqrt{2} \cdot U \cdot sen\omega t \tag{5.12}$$

$$i = \sqrt{2} \cdot I \cdot sen(\omega t - \pi/2) \tag{5.13}$$

Similarly, only by multiplying the voltage at any given instant and the current provides the power at that instant. Then, we plot the instantaneous power curve in this circuit (Fig. 5.3).

Now, let us clarify what this means. As the power is:

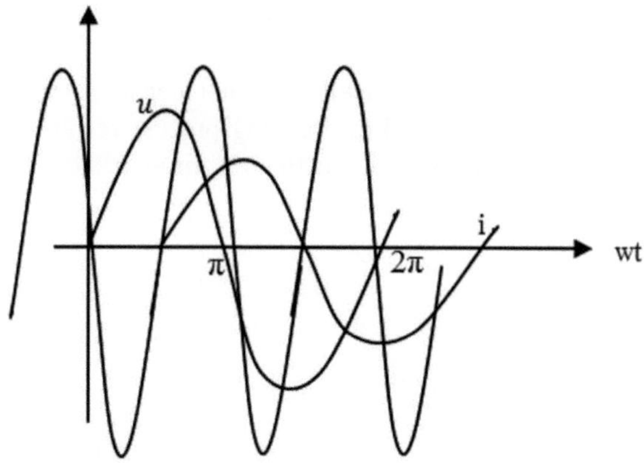

Fig. 5.3 Power curve in a purely inductive circuit. *Source* own elaboration

$$
\begin{aligned}
p(t) &= u(t) \cdot i(t) = \sqrt{2} \cdot U \cdot sen\omega t \cdot \sqrt{2} \cdot I \cdot sen\left(\omega t - \frac{\pi}{2}\right) \\
&= 2 \cdot U \cdot I \cdot sen\omega t \cdot sen(\omega t - \pi/2) = -2 \cdot U \cdot I \cdot sen\omega t \cdot cos(\omega t) \\
&= -U \cdot I \cdot sen2\omega t
\end{aligned}
\tag{5.14}
$$

Recalling the trigonometric functions:

$$
\begin{aligned}
sen2x &= 2senx\cos x \\
cos2x &= cos^2x - sen^2x \\
tg2x &= \frac{2tgx}{1 - tg^2x}
\end{aligned}
$$

If we look at the results, we see that energy has twice frequency that of the original. (Remember: w=2·π·f).

Question What is the behavior if P > 0?

If $p > 0$, then the inductance L acts as a receiver (absorbs and stores energy).

Question What is the behavior when P < 0?

If $p < 0$, inductance L acts as a generator (it returns the energy to the circuit). Only by analyzing the graphical representation we see that the power is positive during the quarter cycle in which the coil is charged (the coil receives energy from the generator and acts as a receiver). In the next quarter cycle, the coil releases its stored energy back to the generator, making the instantaneous power negative. Because the average value of the power during a cycle is zero, it is accurate to state that a pure inductive circuit generates no active power.

Question What is the average circuit power in a pure inductive circuit?

Due to the fact that the mean value of the power is null, a pure inductive circuit does not generate real power.

But, is there power during the coil´s operation if it does not consume any actual energy for its operation?

Although the coil itself does not consume energy during the operation, the constant charging and discharging process results in current flowing through the wires. Power's constant fluctuation through the conductors is known as reactive power (Q).

$$P = -\frac{1}{T}\int_0^T U\, I\, sen\, 2wt = 0 \tag{5.15}$$

In a coil we have that:

$$Q = U_L \cdot I = X_L \cdot I^2 \tag{5.16}$$

We measure it in reactive volt-amperes (VAr).

5.4 Power in a Purely Capacitive Circuit

In this case the current that flows through the circuit leads the voltage π/2 radians (Fig. 5.4). Therefore, the instantaneous values of current and voltage are:

$$u = \sqrt{2} \cdot U \cdot sen\, \omega t \tag{5.17}$$

$$i = \sqrt{2} \cdot I \cdot sen(\omega t + \pi/2) \tag{5.18}$$

In this instance, in order to find the instantaneous power, we multiply the instantaneous voltage and current as we see (Eq. 5.19).

$$p(t) = u(t) \cdot i(t) = 2 \cdot U \cdot I \cdot sen\omega t \cdot sen\left(\omega t + \frac{\pi}{2}\right) = U \cdot I \cdot sen2\omega t \tag{5.19}$$

Then, the active power has twice the frequency of voltage and current waves.

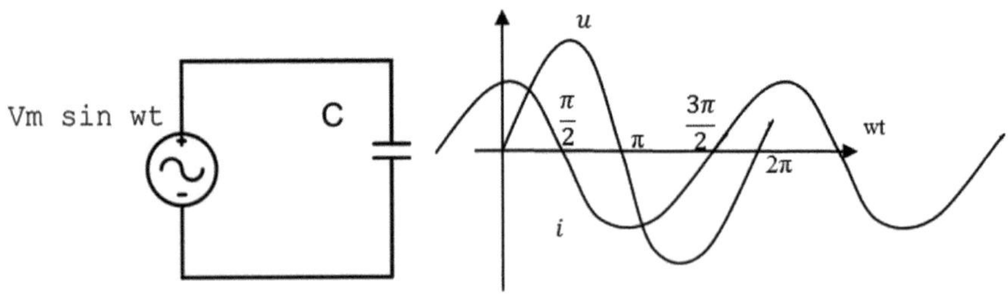

Fig. 5.4 Purely capacitive circuit. *Source* own elaboration

Recalling the trigonometric functions:

$$\begin{aligned} sin(\frac{\pi}{2} + \theta) &= cos\theta \\ sin2x &= 2sinxcosx \\ cos2x &= cos^2x - sin^2x \\ tg2x &= \frac{2tgx}{1 - tg^2x} \end{aligned}$$

As previously, we get the power curve. Once more, the capacitor´s continuous charge and discharge process generate positive and negative cycles of instantaneous power. Then, we can plot the instantaneous power values in a purely capacitive circuit (Fig. 5.5).

Question What happens if p > 0?
If p > 0, then capacitor stores energy.

Question What happens if p < 0?
If p < 0, capacitor releases energy.
What kind of power does a capacitor generate?
A capacitor does not generate active power but reactive power.
What is the active power?

Question The active power is:

$$\begin{aligned} P &= \frac{1}{T}\int_0^T U \cdot I \cdot sen2\omega t = \frac{1}{T}\int_0^T U \cdot I \cdot 2 \cdot sen\omega t \cdot cos\omega t \\ &= \frac{2 \cdot U \cdot I}{T}\int_0^T sen\omega t \cdot cos\omega t = 0 \end{aligned} \tag{5.20}$$

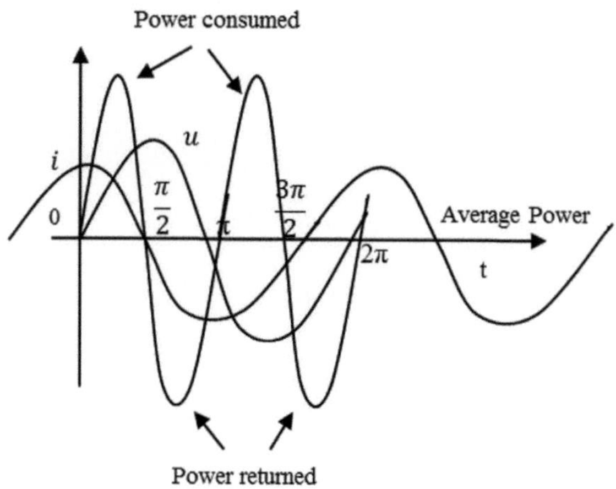

Fig. 5.5 AC power waveforms in a purely capacitive circuit. *Source* own elaboration

Remembering the trigonometric functions of:

$$sin2x = 2sinx \cdot cosx$$
$$cos2x = cos^2x - sen^2x$$
$$tan2x = \frac{2tanx}{1 - tan^2x}$$

If we compare the instantaneous power curves of the coil and the capacitor, we see that during the coil's energy discharge (negative power cycle), the capacitor charges with the same energy (positive power cycle). Consequently, we say that the reactive power of the capacitor is negative with respect to that of the coil, and thus their effects cancel each other out.

In a capacitor we have:

$$Q = U \cdot C \cdot I = X \cdot C \cdot I^2 \tag{5.21}$$

We measure it in reactive volt-amperes (VAr).

5.5 Power in RL Series Circuit

Question What do we mean by an RL circuit?

When we refer to a circuit with R and L, it means that it can either be a circuit with these elements alone or one containing all three receiver components: C, R, and L. The latter implies that the voltage drop across the coil is greater than that across the capacitor, i.e., the inductive effect is more important than the capacitive effect. Both types of power appear in an R-L circuit.

We know that current delays φ degrees, so that:

$$u = \sqrt{2} \cdot U \cdot sen\omega t \tag{5.22}$$

$$i = \sqrt{2} \cdot I \cdot sen(\omega t - \varphi) \tag{5.23}$$

We have that the instantaneous power is:

$$\begin{aligned} p(t) &= u(t) \cdot i(t) = 2 \cdot U \cdot I \cdot sin\omega t \cdot sin(\omega t - \varphi) \\ &= U \cdot I \cdot \cos\varphi - U \cdot I \cdot cos(2\omega - \varphi) \end{aligned} \tag{5.24}$$

Recalling the trigonometric functions:

$$2sin\,A\,cosB = sin(A + B) + sin(A - B)$$
$$2cos\,A\,cosB = cos(A + B) + cos(A - B)$$

$$2sin\,A\,senB = cos(A - B) - cos(A + B)$$

We can build the power triangle of the power wave (Fig. 5.7). The mean value of power indicates the value of the active power:

$$P = \frac{1}{T}\int_0^T p(t) \cdot dt = U \cdot I \cdot \cos\varphi \tag{5.25}$$

Integrating this equation, we have two separate integrals:

$$\begin{aligned}
P &= \frac{1}{T}\int_0^T [U \cdot I \cdot cos\varphi - U \cdot I \cdot cos(2wt - \varphi)] \cdot dt \\
&= \frac{1}{T}\int_0^T \cdot U \cdot I \cdot cos\varphi dt - \frac{1}{T}\int_0^T U \cdot I \cdot cos(2wt - \varphi) \cdot dt = I_1 - I_2 \\
I_1 &= \frac{1}{T}\int_0^T \cdot U \cdot I \cdot cos\varphi dt = \frac{1}{T} \cdot U \cdot I \cdot cos\varphi \cdot \int_0^T dt = \frac{1}{T} \cdot U \cdot I \cdot cos\varphi \cdot [T - 0] \\
I_2 &= \frac{1}{T}\int_0^T U \cdot I \cdot cos(2wt - \varphi) \cdot dt = \frac{1}{T} \cdot U \cdot I \cdot \frac{1}{2w} \cdot [sin(2wt - \varphi)]_0^T \\
&= \frac{1}{T} \cdot U \cdot I \cdot \frac{1}{2w} \cdot [sin(2wt - \varphi)]_0^T \\
&= \frac{1}{T} \cdot U \cdot I \cdot \frac{1}{2w} \cdot [sin(2wt - \varphi) - sin(0 - \varphi)] \\
&= \frac{1}{T} \cdot U \cdot I \cdot \frac{1}{2w} \cdot [sin(2wt - \varphi) - sin(-\varphi)] = 0
\end{aligned}$$

We know that:

$$\begin{aligned}
&sin\,2wt = 0 \rightarrow sin(2wt - \varphi) = sin(-\varphi) \\
&P = I_1 - I_2 = U \cdot I \cdot cos\varphi - 0
\end{aligned}$$

According to these equations, we can plot the vector diagram of the R-L series circuit (Fig. 5.6).

Applying Kirchhoff second law to the circuit, we have that Fig. 5.6.

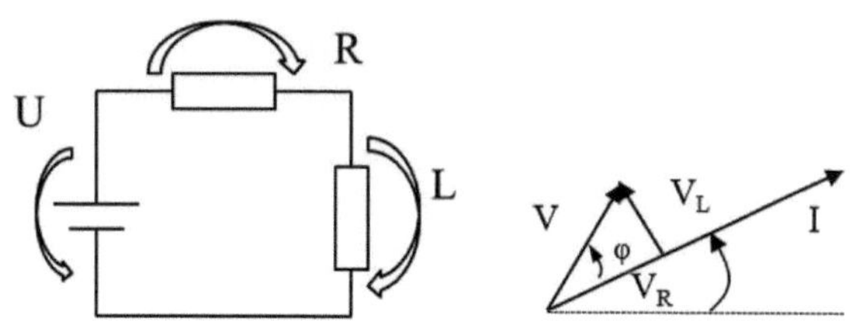

Fig. 5.6 Vector diagram of a R-L Circuit. *Source* own elaboration [2]

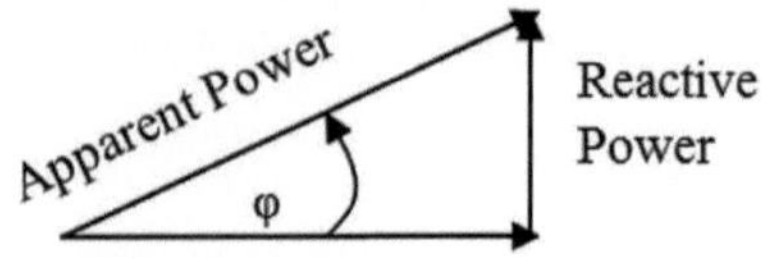

Fig. 5.7 Power-triangle in a R-L circuit. *Source* own elaboration

$$\vec{U} = \vec{U}_R + \vec{U}_L \tag{5.26}$$

We get the power triangle by multiplying the voltage and current in the vector diagram (Fig. 5.7).

$$\vec{U} \cdot I = \vec{U}_R \cdot I + \vec{U}_L \cdot I \tag{5.27}$$

Where we observe that:

$$U_R \cdot I = U \cdot I \cdot cos\,\varphi \tag{5.28}$$

This is the value of the active power. For practical purposes, we consider the coil´s power to be negligible, as we have previously stated the active power of a coil. Then, we consider the power, denoted as S in these diagrams, as decomposable.

$$\vec{S} = \vec{P} + \vec{Q} \tag{5.29}$$

In conclusion, active power produces efficient work in the loads, which includes ohmic resistors, motors, and other passive components.

$$P = R \cdot I^2 \tag{5.30}$$

Or:

$$P = U_R \cdot I \tag{5.31}$$

And generally:

$$P = U \cdot I \cdot cos\varphi \tag{5.32}$$

Reactive power refers to the power that coils and capacitors generate, but the receiver cannot convert into work. The reactive power fluctuates between the generator and receiver, causing a current to flow through the line. This provokes:

$$Q = X \cdot I^2 \tag{5.33}$$

Or:

$$Q = U_X \cdot I \tag{5.34}$$

And generally, we say:

$$Q = U \cdot I \cdot sen\varphi \tag{5.35}$$

$$\vec{S} = \vec{P} + \vec{Q} \tag{5.36}$$

In this case, we have that:

$$S = U \cdot I \tag{5.37}$$

We measure it in volt-amperes (VA).

$$S = \sqrt{P^2 + Q^2} \tag{5.38}$$

Question What does the power factor mean?

The Power factor (PF) in an electrical circuit containing coils and capacitors represents the ratio of active power to total apparent power. We express it as

$$PF = \frac{P}{S} = cos\varphi \tag{5.39}$$

The power factor is the cosine of the phase angle between voltage and current in the power triangle. It's clear that the receiver does not use the total current to generate active power when we look at the applied voltage and current through the circuit diagram.

The remaining portion flows through the wires without contributing to work production. Then,

$$I_a = I\ cos\varphi \tag{5.40}$$

$$I_r = I\ sen\varphi \tag{5.41}$$

Question What does the power factor represent?

It shows that the load only uses a portion of the current. The current comprises two components: the active component of the current (I_a) and the reactive component of the current (I_r). We get the power triangle by multiplying the current triangle and the voltage.

5.6 Summary

This chapter focuses on the concept of electrical power in alternating current (AC) circuits. It begins by defining what electrical power is in an AC context. The chapter then methodically analyzes how power is calculated and behaves in different types of basic circuits: first in a purely resistive circuit, then in circuits containing inductance (a coil) and capacitance individually. Finally, it brings these components together to explain power in a more complex RL series circuit, laying the groundwork for understanding real-world power consumption where resistance and inductance are often combined.

References

1. Omegatron, English: Diagram showing a voltage waveform and the corresponding power waveform (dissipated by a resistive load) (2016). https://commons.wikimedia.org/wiki/File:RMS_voltage_average_power.svg. Accessed 10 Dec 2023
2. Wikipedia. Archivo:SerieRL.png—Wikipedia, la enciclopedia libre». https://commons.wikimedia.org/wiki/File:SerieRL.png. Accessed 10 Dec 2023

Chapter 6
Basic Principles of Electrical Machines

Abstract This chapter describes and classifies electrical machines as well the values of sinusoidal alternating current. In addition, it presents the magnetization curve of ferromagnetic materials as well as the hysteresis cycle. This chapter provides some examples for illustrating the electromagnetic laws to ferromagnetic materials and applies the basic laws to magnetic circuits. It also reviews the magnetic quantities and units as an example of magnetic field generation in electromagnets. It finishes with an analysis of how an electric machine works.

Keywords Electric machines · Static and rotating machines · Magnetic fields · Lenz's law · Hopkinson's law · Barlow's law · Electromagnets · Ferromagnetic materials · Electrical circuit of electric machines · Operation of electric machines

6.1 Classification of Electrical Machines

We classify electrical machines as static or rotating based on their ability to rotate.

6.2 Static Machines: Types

Static machines are those in which all parts do not move during operation. This includes the transformer which we see below Fig. 6.1.

The basic static machine is the transformer. We classify transformers according to the function they perform.

Their task is to transmit electrical power by varying the electrical conditions of the power. A power transformer maintains a constant frequency value while varying the voltage and current values associated with the electrical power it transmits.

Static machines.

1. Transformer
2. Power transformers

D. Borge-Diez et al., *Fundamentals of Electrical Power Engineering*, Lecture Notes in Electrical Engineering 1540, https://doi.org/10.1007/978-3-032-18904-2_6

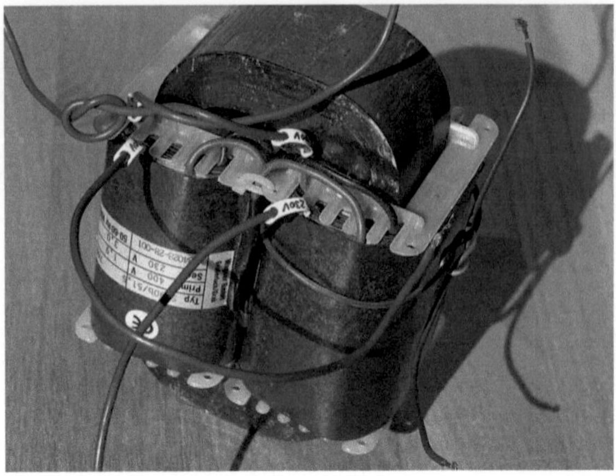

Fig. 6.1 Electric transformer. *Source* From wikicommons under free to use license [1]

3. Special transformers
4. Auto-transformer
5. Measurement transformers
6. Protection transformers
7. Special transformers.

6.2.1 Autotransformer

This term refers to different types of transformers that are used in special applications in the industry and in the grid. The autotransformer is a particularly significant one due to its unique design features. The autotransformer works similarly to as a power transformer, but with specific characteristics that can be drawn (Fig. 6.2).

6.2.2 Measurement Transformers

We use it to measure basic quantities like current, voltage, power, power factor, frequency and synchronization. We find current transformers under different shapes (Fig. 6.3).

6.2.3 Instrument Transformers

They are designed and built to generate an output signal that is proportional and lower than the input signal (Fig. 6.4).

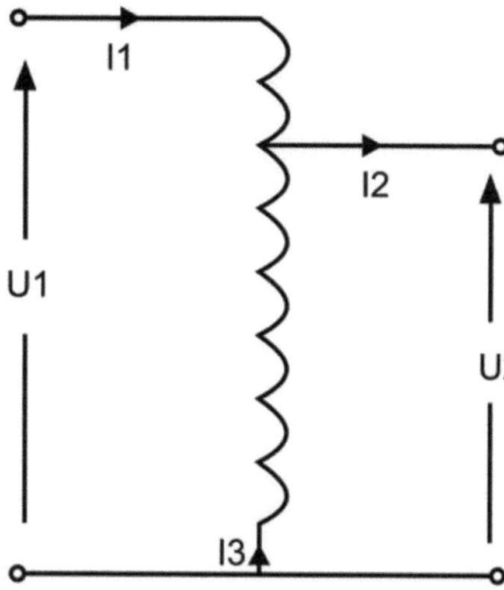

Fig. 6.2 Autotransformer. *Source* From wikicommons under CC0 1.0 use license [2]

Fig. 6.3 Measurement transformers. *Source* From wikicommons under free to use license [3]

Fig. 6.4 Instrument transformer. Original *Source* From wikicommons under CC 3.0 use license [4]

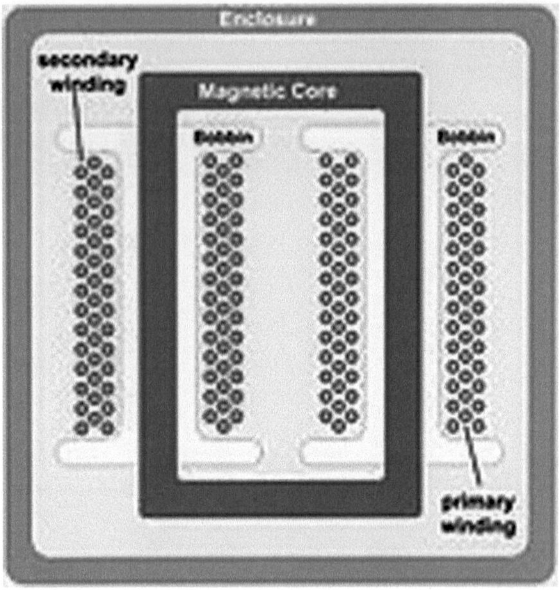

Fig. 6.5 Isolation transformer. *Source* From wikicommons under CC0 1.0 use license [5]

6.2.4 Protection Transformers

Finally, the protection transformers are like instrument transformers, although their task is to provide electrical protection. It is of interest to see the cross section of a simple isolation transformer with no special features (Fig. 6.5). We see the primary, secondary, and electrostatic shield winding, as well as the bobbin, bright, dielectric barrier, magnetic core, and enclosure.

6.3 Static Machines: Active Elements

In general terms, active parts highlight in static machines, specially the windings and the components of a toroid-core transformer (Fig. 6.6). It highlights the primary and secondary windings, as well as the transformer core.

6.3.1 Primary Electrical Circuit

The primary coil has N_1 turns, and receives the power or input signal to the transformer.

Fig. 6.6 Both circuits in a transformer. *Source* From wikicommons under free to use license [6]

6.3.2 *Secondary Electrical Circuit*

The secondary coil has N_2 turns, and sends the power or output signal from the transformer.

6.3.3 *Magnetic Core*

This comprises a material that has high permeability whose function is to provide the magnetic circuit that connects the primary and the secondary.

6.4 Rotating Machines: Active Elements

These machines have fixed but also moving parts. The stator is the fixed part, and the rotor is the part of the machine that moves, mainly rotates. A rotating machine can be arranged used as a motor, or for electricity generation (Fig. 6.7). Among others, the basic components are the stator as well the rotor.

We classify rotating machines according to the function they perform:

Generators

Their task is to generate electricity from other primary energy sources that often have been converted into mechanical energy.

Motors

They convert electrical energy into mechanical energy. The mechanical energy as torque is used to generate work. This torque turns a shaft, allowing a specific amount of work to be done. In addition, it is common to divide rotating machines into two types: DC machines and AC machines. AC machines can function either synchronously or asynchronously. Then we can classify them (Fig. 6.8).

Fig. 6.7 Parts of a rotating machine. *Source* From wikicommons under free to use license [7]

- Generators
  - Direct Current (DC): Dynamos
  - Alternating Current (AC)
    - Synchronous
    - Asynchronous
- Motors
  - Direct Current (DC)
  - Alternating Current (AC)
    - Synchronous
    - Asynchronous
    - Delgas Collector
- Other
  - Stepper motor
  - Etc.

Fig. 6.8 Types of rotating machines. *Source* Own elaboration

6.5 Constituent Elements of Rotating Machines

At least all rotating machines include the following parts:

1. Rotor
2. Stator
3. Air gap
4. Electrical windings.

Definition (*Rotor*) This is the moving part of the machine and comprises a cylinder of magnetic material.

Definition (*Stator*) This is the fixed part of the machine. It consists of a cylindrical crown of magnetic material. It is the stationary part of a rotating machine within which a rotor turns.

Definition (*Air gap*) This is the gap that exists between the rotor and the stator. It is extremely thin when compared to the rotor and stator diameters.

Definition (*Electrical windings*) They are on the outer cylindrical surface of the rotor and the inner cylindrical surface of the stator.

6.6 Energy Conversion in Electrical Machines

The most remarkable active components of the electrical power system are static and rotating AC electrical machines. The energy conversion process that takes place in power transformers occurs by means of the magnetic field (Fig. 6.9). The magnetic field through the magnetic core allows the change between A and B conditions.

This kind of machine requires a magnetic field to transmit and convert high-voltage, low-current energy into low-voltage, high-current energy, and vice versa. A magnetic field makes it possible to convert mechanical energy into electrical energy (Fig. 6.9). Generators require a primary energy source (such as, for example, water moving through a turbine). Then, kinetic energy transforms into mechanical energy. This mechanical energy turns the turbine which drives the generator´s rotor. In electricity generation is common the coupling of a Kaplan water turbine and an electric generator to convert the energy that the water contains (Fig. 6.10).

On the contrary, the energy conversion process in a motor is the reverse of that described for generators. This machine uses the magnetic field to convert electrical energy from the grid into mechanical energy. This mechanical energy is what drives the rotor of the machine.

6.6.1 Need for a Magnetic Field to Operate the Machine

The above shows that a magnetic field is necessary for the operation of any kind of electrical machine, regardless of the kind under consideration. With a battery on

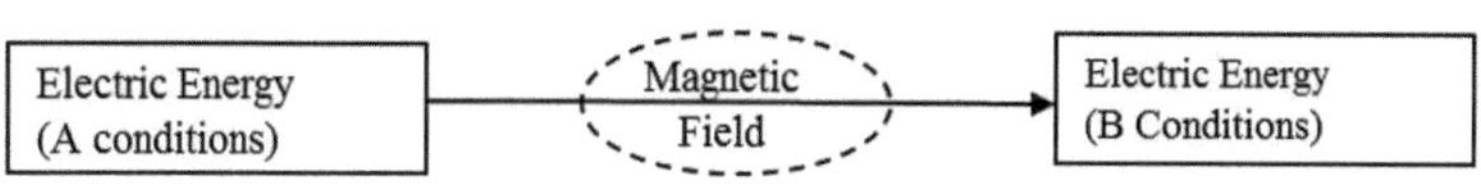

Fig. 6.9 Energy conversion in power transformers. *Source* Own elaboration

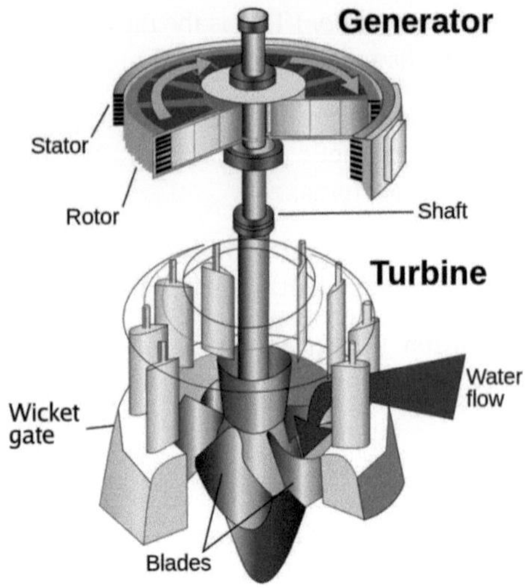

Fig. 6.10 Generator coupled to a Kaplan turbine. Source: PD-US Army. *Source* From wikicommons under free to use license [8]

the left, an iron ring in the middle and a galvanometer on the right, we can arrange Faraday's experiment to induce current from a magnetic field (Fig. 6.11). After the voltage source closes the circuit, the magnetic lines of force move.

Therefore, there are two parts in any electrical machine: one that creates the magnetic field and the other that suffer the effects of it. For instance, the primary in a transformer is the component that creates the magnetic field. In a synchronous machine (Figs. 6.12 and 6.13) the rotor generates the magnetic field and induces the voltage that appears in the stator coils. The stator in an asynchronous machine generates the magnetic field. In transformers these parts are the primary and secondary, while in generators and motors we call them inductor and armature.

It is useful to review the relationship between the component and the function it performs in the energy conversion process (Table 6.1).

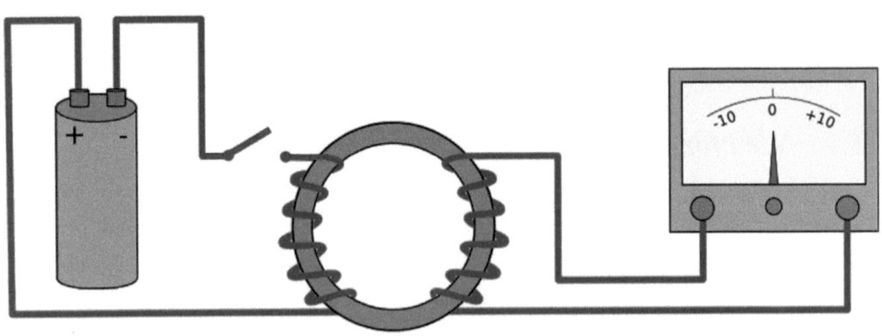

Fig. 6.11 Magnetic circuit. *Source* From wikicommons under free to use license [9]

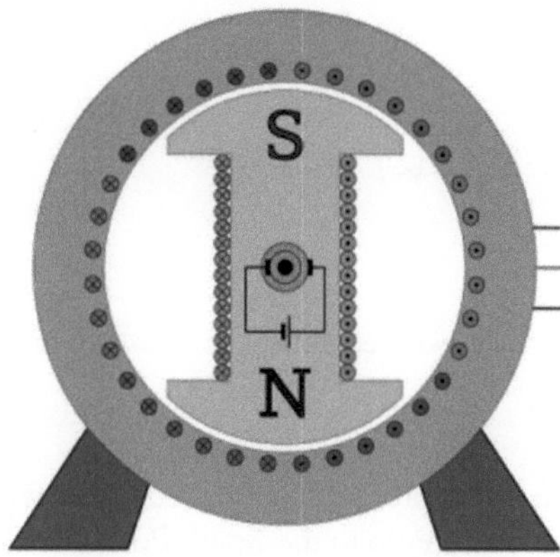

Fig. 6.12 Three Stator windings and two-pole rotor synchronous machine. *Source* From wikicommons under free to use license [10]

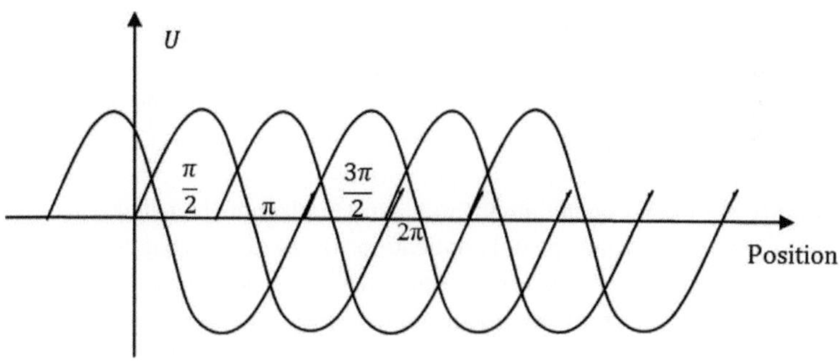

Fig. 6.13 3-Phase voltages diagram. *Source* Own elaboration

Table 6.1 Names of components of electrical machines according to their relation to the magnetic field

	Creates the magnetic field	Receives magnetic field influence
Transformer	Primary	Secondary
Generator	Inductor	Induced
Motor	Inductor	Induced

Source Own elaboration

6.7 Generation of Magnetic Fields

Magnetic fields usually appear from wire current, wire loop current, solenoid current, and earth current (Fig. 6.14).

There is always a magnetic field surrounding any moving electric charge. In some ways, this field is like the gravitational field. We define them as a way of acting on

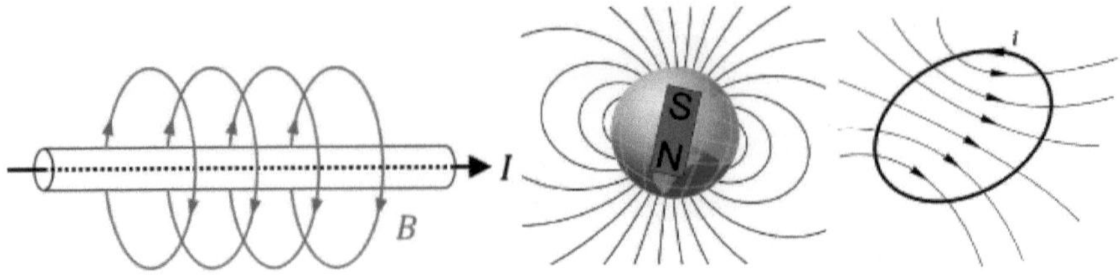

Fig. 6.14 Sources of magnetic fields: wire, earth, current. *Source* From wikicommons under free to use license [11–13]

a charge or a body in space. We represent the magnetic field with force lines. These lines always flow from the North Pole to the South Pole outside of the magnet and from the South Pole to the North Pole inside (Fig. 6.15).

Natural magnets or electromagnets generate magnetic fields. A natural magnet is an object that is surrounded by a magnetic field and that has the property of attracting iron, steel and, to a lesser extent, some other bodies. Strong shocks or a rise in temperature diminish these magnetic properties, which we see in natural magnets. Sometimes, they even vanish over time.

As a result, they are not very flexible when it comes to fitting the intended performance. For this reason, the magnetic fields that natural magnets generate lack the level of controllability that industrial applications often require. As a result, we do not commonly use them to generate the magnetic fields that electrical machines need for operation, except in some special applications. In an electromagnet, we have a set of sheets of magnetic material where a conductor is wound to form a coil of N loops (Fig. 6.16). An electric current passing through a winding produces an electromagnet.

Fig. 6.15 Magnetic field lines of force. *Source* From wikicommons under free to use license [14]

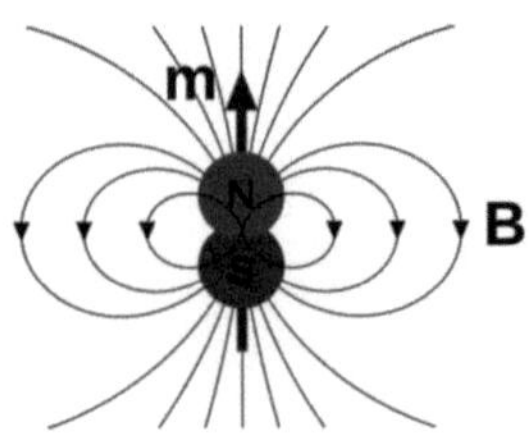

Fig. 6.16 Electromagnet. *Source* From wikicommons under free to use license [15]

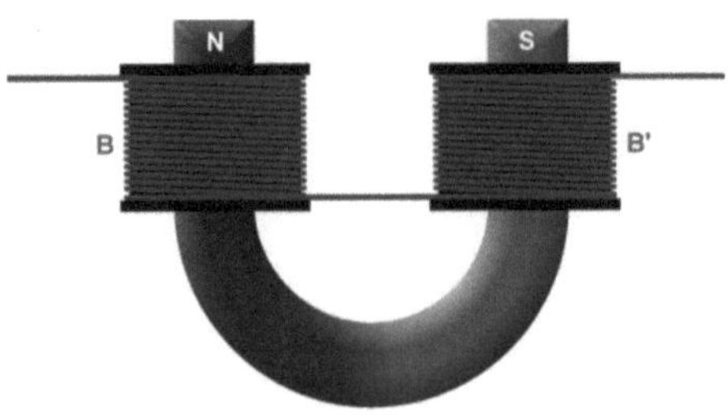

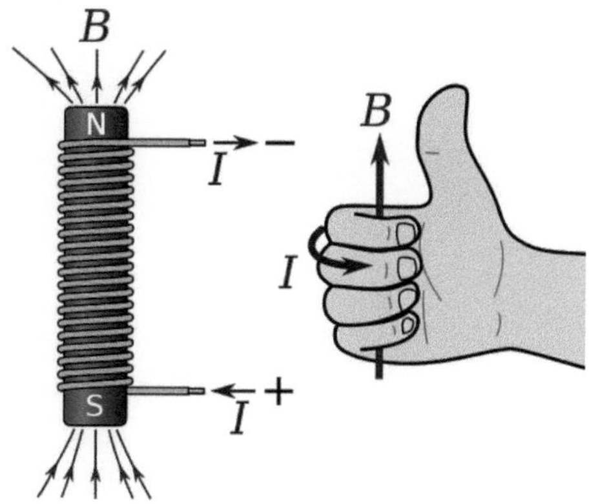

Fig. 6.17 Magnetic field created by the electromagnet. *Source* From wikicommons under free to use license [16]

The current flowing through the coils creates an electromagnet whose magnetic characteristics depend on the current flowing through its coils. This gives electromagnets a high controllability. This is due to the fact that controlling the current is adequate to control the properties of the magnetic field the electromagnet produces on demand. Right hand rule shows the positions of the North Pole and the South Pole for the direction of current flow (Fig. 6.17).

We hold an electromagnet in the right hand, with your thumb extended and the tips of the other four fingers pointing in the direction of the current flowing through the wires. Now, the tip of the thumb indicates the direction of the magnetic lines of force. These lines of force also show the North Pole's position. This is because the lines of force go out at the North Pole and come in at the South Pole. This is the right-hand rule. Remember that changing the direction of current flow also changes the pole's position. Consequently, when we excite the electromagnet with direct current, the position of the poles remains constant in time.

Conversely, when an alternating current excites an electromagnet, the poles' positions shift each time the current's direction changes. Naturally, winding the conductor in the opposite direction on the magnetic material can also reverse the poles' position for the same direction of current flow through the wire. Stated differently, the construction of the electromagnet involves reversing the direction of rotation of the coils. Designed by Belgian inventor Zénobe Gramme in 1871, a Gramme machine was used to generate power (Fig. 6.18). His innovation was to use many armature windings wound on a doughnut-shaped armature and switched with many segmented commutators to smooth the output waveform, generating constant DC power.

In conclusion, because of their controllability, electromagnets are the means by which electrical machines create the magnetic field that they need for their operation.

Fig. 6.18 Gramme machine. *Source* From wikicommons under free to use license [17]

6.8 Elements of Electrical Machines

(a) Magnetic circuit

With transformers, this circuit consists only of stacked magnetic plates (Fig. 6.19).

With rotating machines, it also comprises the air gap that locates between the stator and the rotor. Electrical machines are made up of two electrical circuits that are kept apart electrically and connected by magnetism, with the exception of autotransformers.

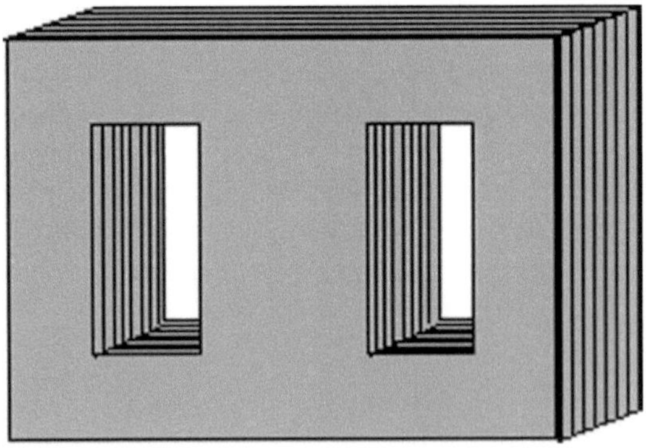

Fig. 6.19 Stacked magnetic plates. *Source* From wikicommons under free to use license [18]

6.8.1 Electric Circuit and Magnetic Circuit

One circuit generates the magnetic field and other the effects of it. Electrical circuits commonly comprise windings, except in special cases such as squirrel-cage induction motors. Transformers windings are wound around the core.

In rotating machines, the loops that locate in the slots form one electrical circuit on the outer circumference of the magnetic cylinder corresponding to the rotor. The coils form the other electrical circuit on the inner circumference of the magnetic circuit corresponding to the stator. Active parts comprise the magnetic circuit and the electrical circuit. The reason for this is that these are the two main components that enable the operation of an electric machine. However, an electric machine must also include several systems that allow the active parts to operate. These systems mainly include the cooling system, the insulation system, the regulation and control system and the protection system.

When an electrical machine operates, several power losses occur in its active parts. We mainly classify them in iron and copper losses. Both the iron losses (corresponding to the magnetic circuit) and the copper losses (corresponding to the electrical circuit) manifest themselves as heat, which raises the machine´s internal temperature.

Then, the purpose of the cooling system is to dissipate the heat that the machine generates. This prevents the temperature in the different parts of the machine components from rising above the upper limit, which they are designed for. In order to encourage the removal of heat to the outside, the cooling system may include a ventilation system that removes the medium in direct contact with the machine casing, commonly air.

6.8.2 Isolating System

It is necessary to isolate the parts of the machine that operate at different voltage levels. Many cases, e.g. transformer oil, the same fluid performs both cooling and insulating functions. The protection system must cover electrical, mechanical, and personal safety conditions.

Electrical protective devices must protect machinery both against external disturbances that may reach it and against internal disturbances resulting from fault or malfunction. In the latter case, they must also prevent the effects of spreading outside the machine. The components of a magnetic circuit are the same as those of an electrical circuit: they are a group of linked parts that allow a magnetic flux to flow through them. There are similarities when modeling electric and magnetic circuit (Fig. 6.20).

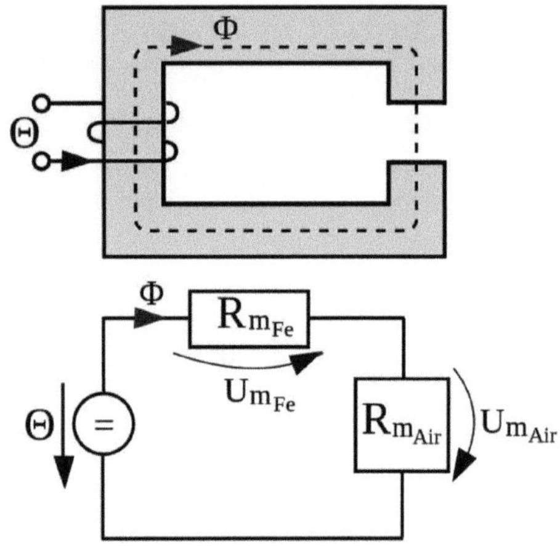

Fig. 6.20 Example of a magnetic circuit to demonstrate voltage divider. *Source* From wikicommons under free to use license [19]

6.9 Magnetic Quantities and Units

Then, the magnitudes in a magnetic circuit (and their corresponding units) are shown now on:

Important Magnetic flux, Φ (Weber, Wb).

Magnetic flux is the set of lines of force passing through a magnetic circuit. These lines of force are always closed on themselves. The unit that we measure magnetic flux in the international system is the Weber (Wb).

Important Magnetic induction, B (Tesla, T).

Magnetic induction, or flux density, is the flux per unit area. This is what happens if the flux is evenly spread across the whole surface:

$$B = \frac{\Phi}{S} \tag{6.1}$$

Important Magnetomotive force (MMF).

It is the work that would be required to carry a hypothetical isolated magnetic pole of unit strength completely around a magnetic circuit. Flux circulates in a magnetic circuit due to the application of magnetic force. A coil having N turns through which current I flows produces a magnetomotive force of:

$$F_{mm} = N \cdot I \tag{6.2}$$

The international system uses the ampere-turn as its unit of measurement (At).

Important Magnetic field strength (H).

Magnetic field strength or magnetizing force is the ratio of the applied magnetomotive force (MMF), in ampere-turns (At), to the length of the magnetic circuit consuming it.

We measure the length on the mean magnetic line, L, as.

$$H = \frac{F_{mm}}{L} \tag{6.3}$$

The unit of measurement in the International System is (ampere-turn)/meter (At/m). The magnetic permeability of vacuum (μ_0) is 4π-10–7 H/m, in the international system. We use ferromagnetic materials to make electrical machines because it is important to keep the magnetic flux inside them.

Important Permeability (μ).

The permeability of a material measures the ability of that material to conduct magnetic flux through it. Conversely, relative permeability (μ_r) quantifies the capacity to conduct flux with respect to vacuum. The relative magnetic permeability of air has a value of one.

Important Reluctance ($\Re$).

The reluctance of a magnetic circuit (in Ampere-turn/Wb) is the resistance of the magnetic circuit to the flow of flux through it. This is similar to that of "resistance" in an electrical circuit. The reluctance in a circuit of length L, constant cross section S and material of permeability μ is defined as:

$$\Re = \frac{L}{\mu \cdot S} \tag{6.4}$$

This expression shows that the higher the value of the permeability of the material, the lower the value of the reluctance and, the less resistance the circuit

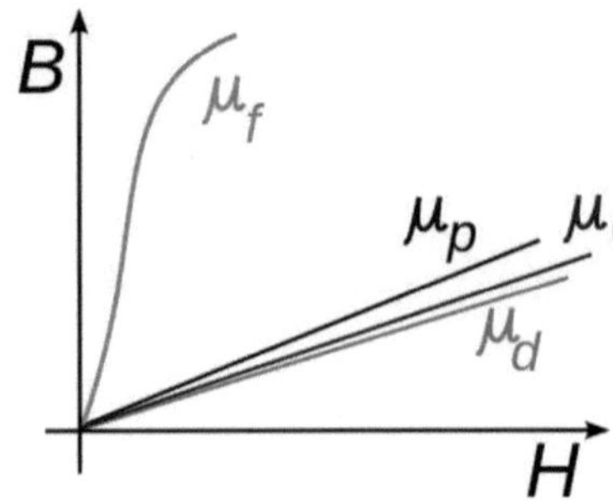

Fig. 6.21 B—flux density, H—magnetic field, μ_f—permeability of ferromagnets (and ferrimagnets), μ_p—permeability of paramagnets, μ_0—permeability of free space ($4\pi\ 10^{-7}$ H/m), μ_d—permeability of diamagnets. *Source* From wikicommons under free to use license [20]

presents to the flux. Materials behave differently according to the flux density and magnetic field (Fig. 6.21). It includes B—flux density, H—magnetic field, μ_f—permeability of ferromagnets (and ferrimagnets), μ_p—permeability of paramagnets, μ_0—permeability of free space ($4\pi\ 10^{-7}$ H/m), μ_d—permeability of diamagnets.

This is the reason for manufacturing electrical machines' magnetic circuits using ferromagnetic materials (logically excluding the air gap in rotating machines).

6.9.1 Basic Laws

Definition (*Hopkinson's Law*) Also known as the induction flux law, it states that flux in a magnetic circuit is directly proportional to the applied magnetomotive force and inversely proportional to circuit reluctance. Then, we can write:

$$M_{MF} = \Phi \cdot \Re \tag{6.5}$$

And substituting Eq. 6.2 we have:

$$\Phi \cdot \Re = N \cdot I \tag{6.6}$$

where $\Re$ is the reluctance in Ampere-turn/Wb, Φ is the magnetic flux in Weber, Wb, and M_{MF} is the magnetomotive force in ampere-turn (At).

Definition (*Lenz's Law*) The flux Φ (t) perpendicular to the surface that a coil with N loops defines induces the following emf (e) at the coil's ends:

$$e(t) = -N \cdot \frac{d\Phi(t)}{dt} \tag{6.7}$$

The induced emf produces current that is opposite to the flux variation that causes it, as indicated by the minus sign.

Question Why does this law support the way the transformer function?

Keep in mind that when we place a coil in a magnetic circuit where alternating current flows, an alternating voltage appears at its ends. When direct current flows through the circuit, then the flux is constant and time-independent. If we place the coil in it, no voltage induction appears (for this reason, there is no direct current transformer). This is the operating principle of the transformer. According to Lenz's law, to induce voltages, flux must be variable (Fig. 6.21).

6.9.2 Electromagnetism in Ferromagnetic Materials

For simplicity, we use in this section a homogeneous series magnetic circuit. This is how we name that a circuit through which the same flux flows (in series) and which consists of a single type of material having a constant area (homogeneous).

6.9.3 Direct Current Excitation

Homogeneous magnetic series circuit.

The circuit below Fig. 6.22 shows an electromagnet with a closed magnetic circuit: This is excited by N windings through which direct current I flow. The magnetic flux Φ circulates as a result of this excitation. The quantities needed to find the magnetic field using Ampere's Law is: l_1 is the length of the flux path or magnetic circuit lying within the magnetic core in meters, H_1 is the strength of the magnetizing field in the core in ampere-loops, l_2 is the length of the air gap, and H_2 is the strength of the magnetizing field in the air gap.

According to Ampere's Law we have: $N \cdot I = H_1 \cdot l_1 + H_2 \cdot l_2$.

It appears the flux of value (Φ) for every value of the current (I). We express these quantities in terms of their corresponding magnetic field strength (H) and induction

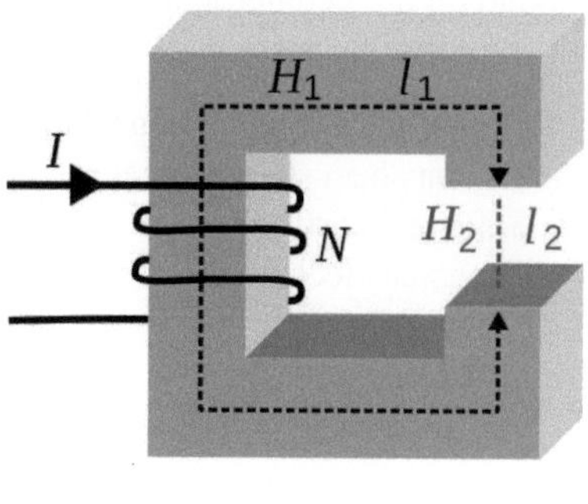

Fig. 6.22 Magnetic circuit series. *Source* From wikicommons under free to use license [21]

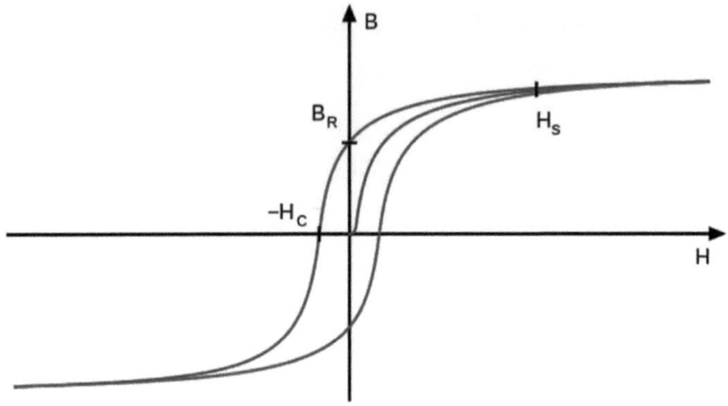

Fig. 6.23 Hysteresis curve and first magnetization. *Source* From wikicommons under free to use license [22]

(B) values.

$$H = \frac{NI}{L} \tag{6.8}$$

$$B = \frac{N\phi}{S} \tag{6.9}$$

Only plotting each pair of values (B, H), we see it corresponds to a point on the coordinate axes (Fig. 6.23). Representing all points (from H = 0 to whatever value is set as the upper limit) gives the so-called magnetization curve or first magnetization curve. Developing the reluctance expression and grouping terms, we can express Hopkinson's law as:

$$B = \mu \cdot H = \mu_0 \cdot \mu_r \cdot H \tag{6.10}$$

The only factor influencing the relationship between B and H is the permeability of the material which we use to build the magnetic circuit. Graphs of the magnetization curve of magnetic materials are very similar (Fig. 6.23).

Therefore, the first magnetization curve is unique to each material and does not depend on the geometric dimensions of the circuit (Fig. 6.24).

The substances are:

1. Standard sheet steel, annealed,
2. Silicon sheet steel, annealed, Si 2.5%,
3. Soft steel casting,
4. Tungsten steel,
5. Magnet steel,
6. Cast iron,
7. Nickel, 99%,

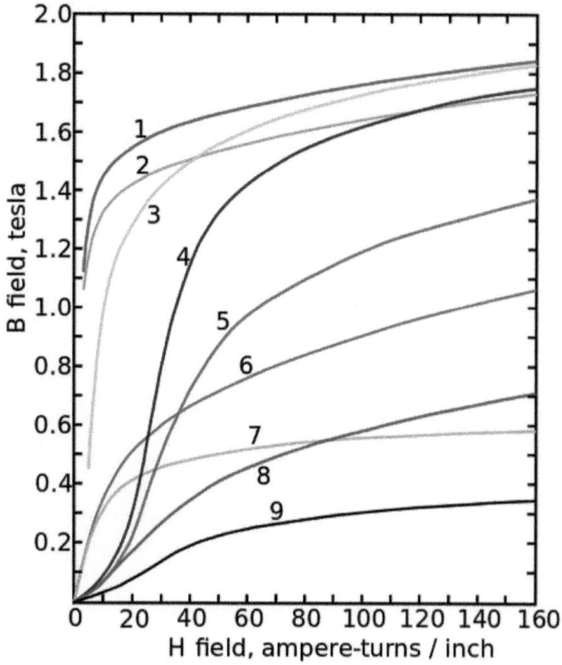

Fig. 6.24 Hysteresis: Tracing of graph downloaded from Charles Steinmetz (1917) Theory and Calculation of Electric Circuits, McGraw-Hill, New York, USA, p. 84, Fig. 42 on Google Books. *Source* From Wikipedia under free to use license [23]

8. Cast cobalt,
9. Magnetite, Fe_2O_3.

Ferromagnetic materials have a very similar magnetization curves (Fig. 6.25). There we see different zones of magnetization.

In ferromagnetic materials, there are three distinct zones that make up the line of first magnetization.

Non-saturation zone

In this first region, we consider that the B-H relationship is linear. As a result, an increase in the magnetic field strength (H) causes an equal increase in induction (B).

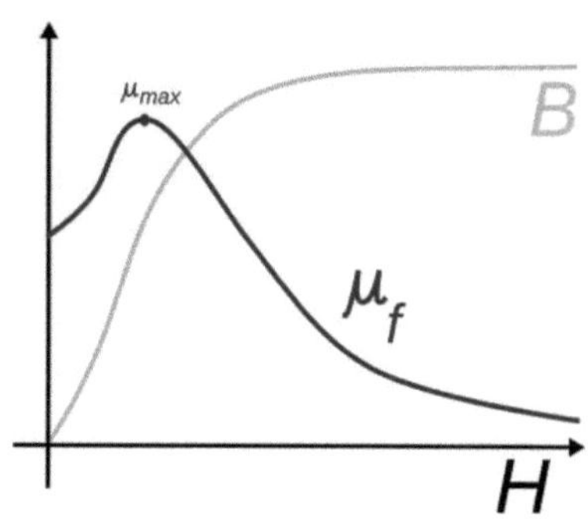

Fig. 6.25 Magnetization curve of ferromagnetic materials. *Source* From wikicommons under free to use license [24]

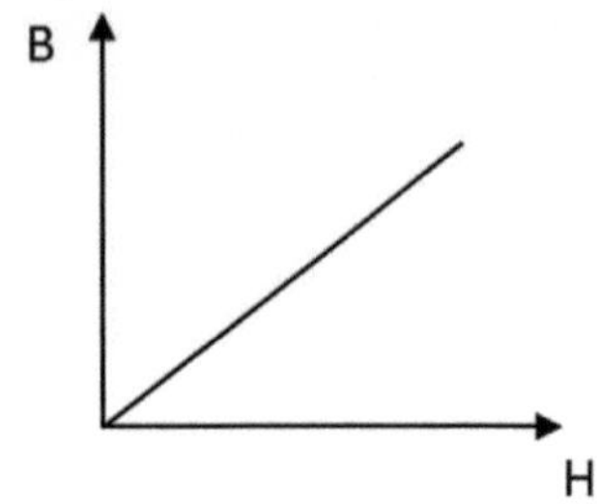

Fig. 6.26 First magnetization curve of air. *Source* Own elaboration

Saturation elbow

This is the area where the B-H ratio loses its linear character and the effects of saturation become apparent.

Saturation zone

Because significant increases in H match minor increases in B, this zone is distinctive. It takes a much larger excitation to achieve a small increase in flow rate.

The magnetization curve shows that the relative permeability of ferromagnetic materials is not constant, and depends on how we use the material. For a wide range of operation we have that:

$$\frac{B}{H} = \mu_0 = cte \tag{6.11}$$

The relative permeability of air is constant and unitary whatever the operating conditions (Fig. 6.26).

For this reason, the first magnetization curve of air is a straight line with slope μ_0. This is because this medium has no saturation. As we know with direct current, the current (both in value and direction) remains invariant. Therefore, the flux is also invariant in direction and of constant value. This means that a pair of B-H values (corresponding to certain operating conditions in the circuit) defines an operating point that is on the first magnetization curve and remains in a fixed position.

6.9.4 Alternating Current Excitation

If an alternating current I (t) passes through the loops, a flux pass through the magnetic circuit (Fig. 6.23). Summarizing, the flux changes with time in both value and direction, which causes a change in the induction as well in the strength of the magnetic field (Fig. 6.26). Then, we have variable quantities that we define as:

$$H(t) = \frac{N \cdot i(t)}{L} \tag{6.12}$$

$$B(t) = \frac{\Phi(t)}{S} \tag{6.13}$$

Question What are the operating conditions of the magnetic circuit?

All of this suggests that, in an alternating current system, the location of the B-H plane's representative point for the operating conditions of the magnetic circuit is not constant because it varies with time B (t)–H (t).

Question In ferromagnetic materials, what is the name of the path that the operating point B (t)–H (t) describes?

We call the hysteresis cycle to the path that the operating point (over one cycle of the alternating excitation current) describes for ferromagnetic materials. The operating point moves counterclockwise through this cycle (Fig. 6.27). There is a unique hysteresis cycle for every maximum excitation current value.

Every maximum magnetic field strength (H_{max}) or corresponding maximum induction (B_{max}) value results in a different hysteresis cycle. We observe that the induction continues even if the excitation stops (H = 0). Here, the material keeps a value: the residual magnetism, which we designate as B = OR, or B = OR′. The magnetic hysteresis curve of a magnetic memory core during a reading operation is also illustrated (Fig. 6.27). Sense line current pulse is high ("1") or low ("0"). Depending on the initial magnetization state of the core, the sense line current pulse is either low ("0") or high ("1").

Question What variables determine if B reaches OR, or OS value?

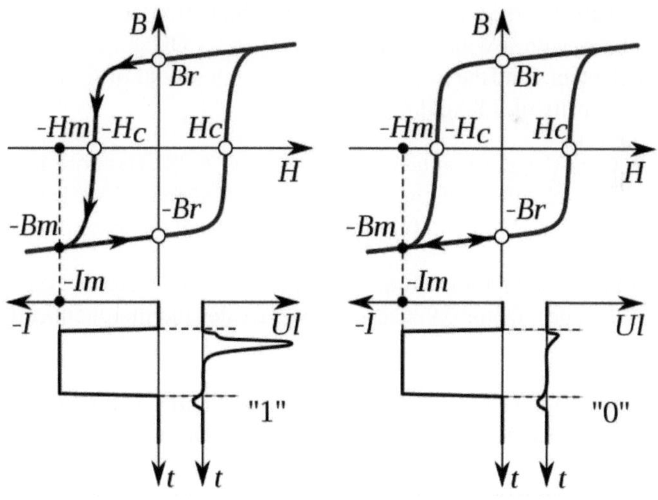

Fig. 6.27 Hysteresis cycle. *Source* From wikicommons under free to use license [25]

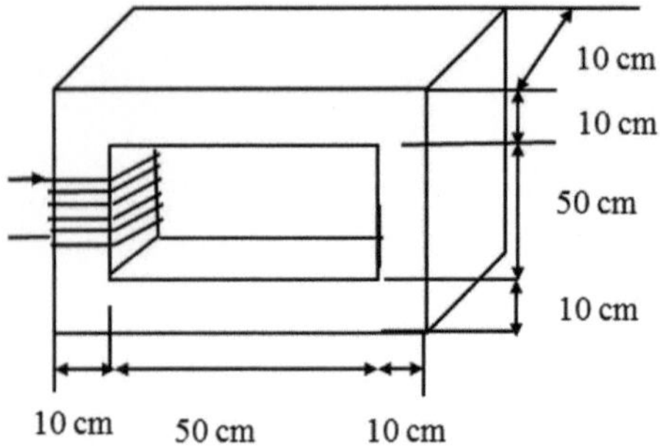

Fig. 6.28 Problem 6.1. *Source* Own elaboration

Determining if OR, or OS value is contingent upon the operating point of the circuit during excitation cancellation. The magnetic field strength (H = OP or H = OQ) that produces the zero value of induction is what we call the coercive force.

We do not observe hysteresis cycle when the medium is air. Therefore, the operating point in alternating current moves along a straight line in time, and at constant slope in the first and third quadrant, μ_0.

Problem 6.1 Suppose we have the magnetic circuit shown below Fig. 6.28, consisting of stacked magnetic sheets of thickness 0.5 mm. Iron losses supposes 2.6 W/kg measured at a frequency of 50 Hz having 1 T of induction.

Determine the number of loops that the circuit needs to achieve 1.5 T of magnetic field strength when a DC current of 75 A flows through the coil.

Compare this problem, but considering an air gap of 1.5 mm.

Assuming there is no air gap; determine the induction at which the magnetic material functions with a winding of 125 loops and a DC current of 40 Amperes flowing through them. Find the iron losses under the conditions of: 50 Hz of frequency and a material density of 7.8 kg/dm^3.

Problem 6.2 Consider the magnetic circuit of Fig. 6.28. The core is made from magnetic sheet metal and insulating material. The magnetic sheet is 0.5 mm thick and the iron loss is 2.6 W/kg at a frequency of 50 Hz and an induction of 1 T.

We know that the fill factor is a function of the insulating material as: Paper: factor 0.88. Varnish: (1 side): factor 0.9. Carlite tape: (2 sides thermo-chemically treated): factor 0.97.

Problem 6.3 Suppose we have an iron core with a winding as Fig. 6.29 shows. The useful cross section of the core is 150 cm^2. The air gap is 0.75 mm. The average length of the iron circuit is 250 cm and it is made of 0.5 mm thick magnetic sheet with iron losses of 2.6 W/kg at a frequency of 50 Hz with induction 1 T.

Find the number of loops N if the direct current flowing through them is 75 A DC and we want to obtain a flux of 0.0195 Wb (Fig. 6.30).

Find the flux if N = 526 and I = 25A.

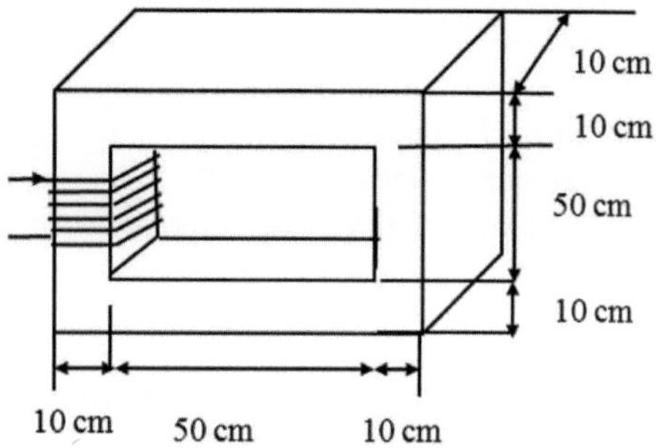

Fig. 6.29 Problem 6.2. *Source* Own elaboration

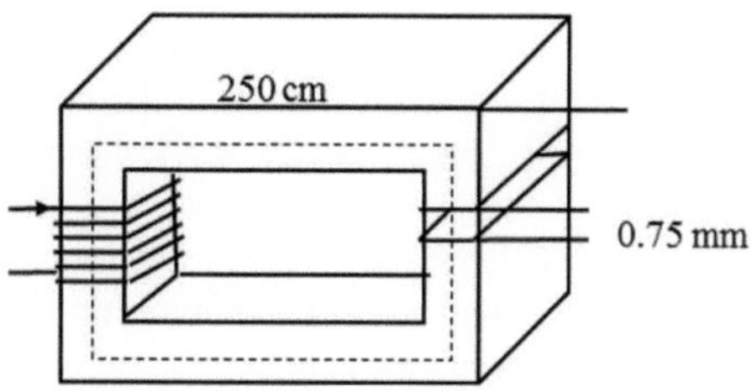

Fig. 6.30 Problem 6.3. *Source* Own elaboration

Problem 6.4 Suppose we have a magnetic circuit that is made of 0.5 mm thick magnetic sheet with iron losses of 2.6 W/kg measured at a frequency of 50 Hz and an induction of 1 T (Fig. 6.31). Assuming a fill factor of 0.95, find the number of loops N so that:

The number of windings N so that $\Phi_3 = 0.0125$ Wb when a direct current of 10 A flows through them.

The iron losses, knowing that the density is 7.8 kg/dm^3.

Hints:

We define the stacking or filling factor as the ratio of the useful section (of magnetic material) and the total section (Fr = Su/St). As the p-flow will only circulate through the useful section, the induction is expressed as:

$$B\frac{\Phi}{S_u} = \frac{\Phi}{F_r \cdot S_{Total}}$$

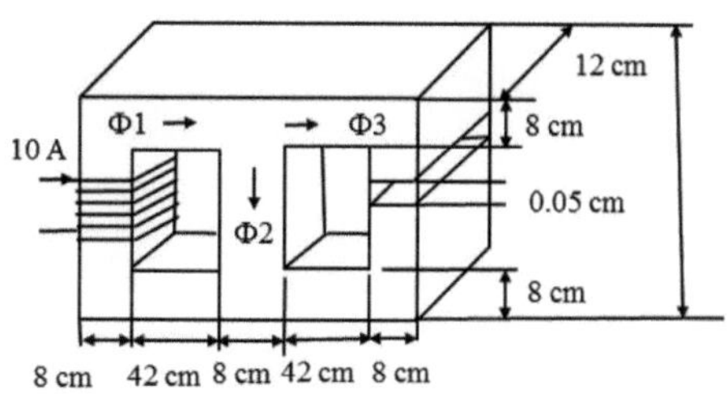

Fig. 6.31 Problem 6.4. *Source* Own elaboration

This value depends on the thickness of the sheet and the type of insulation used. If we use sheet thicknesses of 0.35 mm, typical values of insulation are:

Paper: (1 side): factor 0.88.

Varnish: (1 side): factor 0.9

Carlite material: (2 sides thermo chemically treated): factor 0.97.

6.10 An Application Example of Fundamental Laws: The Barlow Wheel

In this section, we review a case study demonstrating the practical application of electrical machines that we discussed. Peter Barlow invented the Barlow wheel in 1822 to demonstrate the principle of a homopolar electric motor. The star-shaped wheel pivots for rotation and has points that dip into liquid mercury, which is contained within a wooden base. During operation, a horseshoe magnet is positioned on the wooden base, with its poles embracing the wheel. The wheel then rotates when an electric current passes between the hub of the wheel and contacts the mercury.

Now we analyze the Barlow wheel's operation as a motor to ascertain its rotational speed expression.

6.10.1 *Barlow's Law*

The Barlow wheel (Fig. 6.32) comprises a star-shaped wheel, pivoted so it is free to rotate about its axis, with the points dipping into a trough of the liquid metal mercury in the wooden base. In order to work, a horseshoe-shaped magnet with its poles encircling the wheel is set on the wooden base. When an electric current flows between the wheel hub and the mercury contact, the wheel rotates.

Suppose we apply the voltage E between the center and the outside of a metal disc placed inside a magnetic field B. Current i flow through the points (x), which form a radius, and due to the induction B, a force F appears, which makes the disc rotate. An electromotive force that opposes the voltage emerges when the disc rotates:

$$e.m.f. = -\frac{d\varphi}{dt} \tag{6.14}$$

The behavior of the disc is then defined:

The current flowing through the disc is:

$$i = \frac{E - \frac{d\varphi}{dt}}{R} \tag{6.15}$$

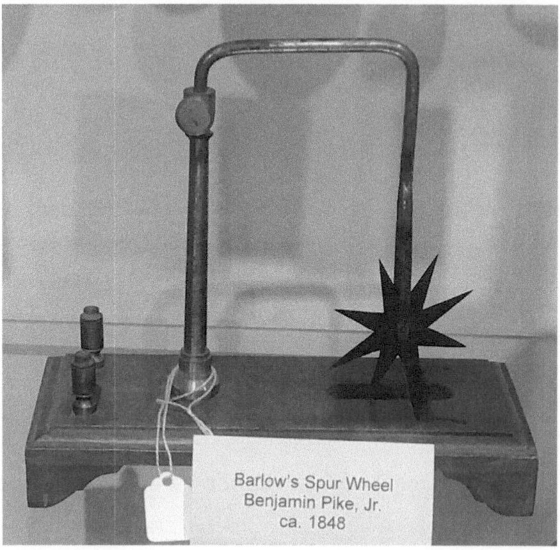

Fig. 6.32 A Barlow´s Wheel. *Source* From wikicommons under free to use license [26]

where R is the electrical resistance of the circuit and I is the closed-loop current flow. Suppose the disc rotates at a speed of n r.p.s. Then, we have:

$$\frac{d\varphi}{dt} = B \cdot S \cdot n \tag{6.16}$$

Substituting Eqs. 6.14 and 6.16 in Eqs. 6.15, and clearing i yields:

$$i = \frac{E - B \cdot S \cdot n}{R} \tag{6.17}$$

Then the force dF appears at the differential of length dx. We express it as:

$$dF = B \cdot i \cdot dx \tag{6.18}$$

The torque that this force creates is:

$$dCm = x \cdot dF = B \cdot i \cdot x \cdot dx \tag{6.19}$$

Therefore, we get the total torque only by integrating from 0 to r:

$$C_m = \frac{Bir^2}{2} = \frac{Br^2}{2}\frac{E - B \cdot S \cdot n}{R} \tag{6.20}$$

Remember the equations of:
J: Moment of inertia of the solid.
C: Torque. $C = J \cdot \alpha$

P: Power. $P = C \cdot w$

Recalling the Newton´s law for rotating systems we have:

$$C_m - C_r = J \cdot \frac{d\omega}{dt} \tag{6.21}$$

where Cm is the torque in N·m; Cr is the resistant torque, in N·m; $w = \frac{d\theta}{dt}$ is the angular speed, in rad/s; $\alpha = \frac{d\omega}{dt}$ is the angular acceleration, in rad/s^2; and J is the moment of inertia, in kg·m^2.

Remember that we define angular acceleration as the ratio of change in angular speed:

$$\alpha = \frac{\Delta w}{\Delta t} \text{Angular acceleration} \left(\text{rad/s}^2\right)$$

From the frequency change, one can also determine the angular acceleration, since w = 2πf.

$$\alpha = \frac{2\pi(\Delta f)}{t}$$

Therefore, we obtain the total torque integrating from 0 to r:

$$C_m = \frac{Br^2}{2}\frac{E - B \cdot S \cdot n}{R} - C_r = J \cdot 2\pi \frac{dn}{dt} \tag{6.22}$$

where C_r is the resistant torque in N·m; J is the moment of inertia, in kg·m^2; and w is the angular acceleration, in rad/s^2.

In general, we know the resistive torque (Cr) and we want to determine speed (n). Since the speed does not change, its time is null. Consequently, for a given resistive torque the expression for speed remains:

$$n = \frac{E - \frac{2 \cdot Cr \cdot R}{B \cdot r^2}}{B \cdot S} \tag{6.23}$$

Question What are the units of n?

$$\begin{aligned} &\frac{V - \frac{N \cdot m \cdot \Omega}{\frac{\Phi}{m^2} \cdot m^2}}{\frac{\Phi}{m^2} \cdot m^2} \to \frac{V \cdot \Phi - N \cdot m \cdot \Omega}{\Phi^2} \to \frac{V^2 \cdot s - N \cdot m \cdot \Omega}{V^2 s^2} \\ &\to \frac{1}{s} - \frac{N \cdot m \cdot \Omega}{V^2 s^2} \to \frac{1}{s} - \frac{N \cdot m \cdot V^2}{V^2 \cdot s^2 W} \to \frac{1}{s} - \frac{W}{sW} \to S^{-1} \end{aligned} \tag{6.24}$$

$$\Omega = \frac{V}{A} = \frac{w}{A^2} = \frac{v^2}{w};$$

And:

$$W = \frac{J}{s} = \frac{N \cdot m}{s} \tag{6.25}$$

6.11 Electrical Circuit of Electrical Machines

Loops make up the electrical circuit of most machines, with a few notable exceptions, such as squirrel cage rotors in asynchronous machines. There are various ways to place these loops. Winding the conductor on the appropriate magnetic core creates the loops in transformers. We can arrange them differently, although it is common to use concentric windings in static electric machines to save material, and space (Fig. 6.33).

The two most widely used constructive arrangements are separate windings (Fig. 6.34) especially in a static electric machine. In rotating machines, the windings comprise coils joining two active sides.

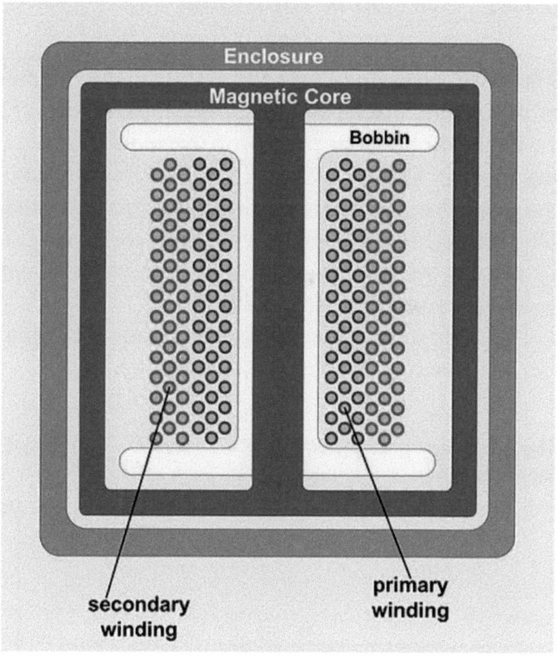

Fig. 6.33 Concentric windings in an isolation transformer. *Source* From wikicommons under free to use license [27]

Fig. 6.34 Separate windings. *Source* From wikicommons under free to use license [28]

6.11.1 Maximum Utilization

Thus, two active sides form a loop (Fig. 6.35), and the winding corresponding to one of the rotating machine's phases is formed by the set of loops.

We place the active sides of the loops that become the windings in rotating machines in slots in the inner perimeter of the stator or in the outer perimeter of the rotor. For maximum efficiency, the space between the active sides of the coils (coil pitch) (Fig. 6.35) should coincide with the space between two consecutive poles of different polarity (polar pitch). We can see the coil pitch as well the rotor windings in a rotating machine (Fig. 6.36).

There are several alternative configurations either a single-phase bipolar machine, i.e. with one pair of poles (p = 1) (Fig. 6.37).

Fig. 6.35 Arrangement of a loop in an electrical machine. *Source* Own elaboration

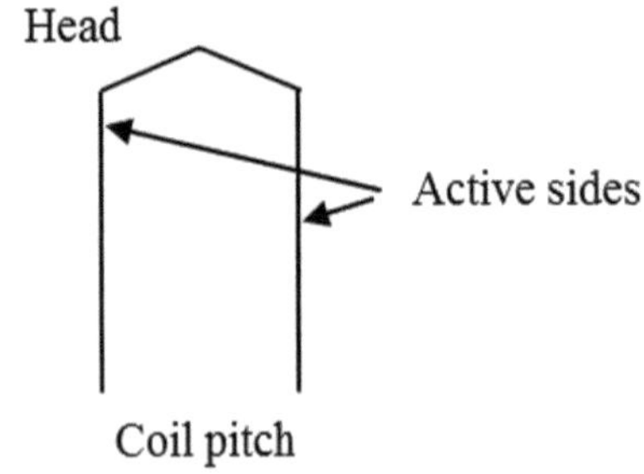

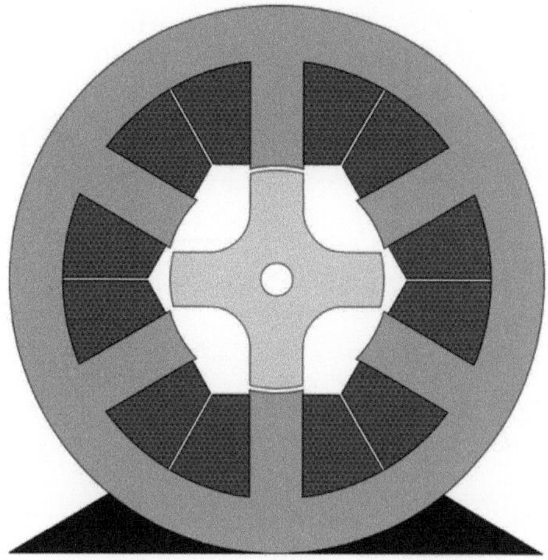

Fig. 6.36 Switched reluctance motor 6–4. *Source* From wikicommons under free to use license [29]

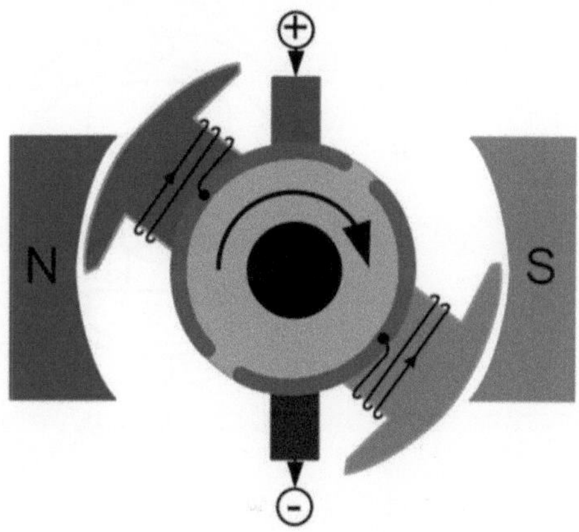

Fig. 6.37 2-pole rotating machine. *Original Source* From wikicommons under CC 3.0 use license [30]

The distance between the rotor's poles and the active sides of the coils that comprise the winding—which are situated at A-AA′ and B-BB′ of the stator—coincides (Fig. 6.38). We can figure out the configuration of a single-phase, four-pole machine with two pairs of poles (p = 2) (Fig. 6.38).

The three-phase windings follow the same layout principles as the single-phase windings in each phase. For example, we can draw a three-phase two-pole machine (p = 1) (Fig. 6.39). We have to remember that there must be a 120° electrical separation between the three phases for a balanced system.

When p = 1, these electrical 120º are also geometrical 120º. Thus, the angle between A1-A1′ and A2-A2′ is 120º and between A2-A2′ and A3-A3′ is also 120º. Depending on several factors, the connection between the windings of the three phases—which can be independent—will be in a star or delta configuration (Fig. 6.39). Other common arrangement is a single-phase, four-pole machine (p = 2) (Fig. 6.40).

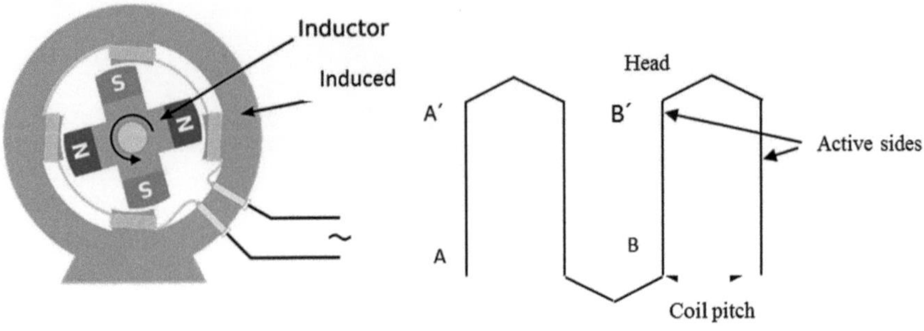

Fig. 6.38 Single-phase, two-pole, two-pole-pair winding. *Source* From wikicommons under free to use license [31]

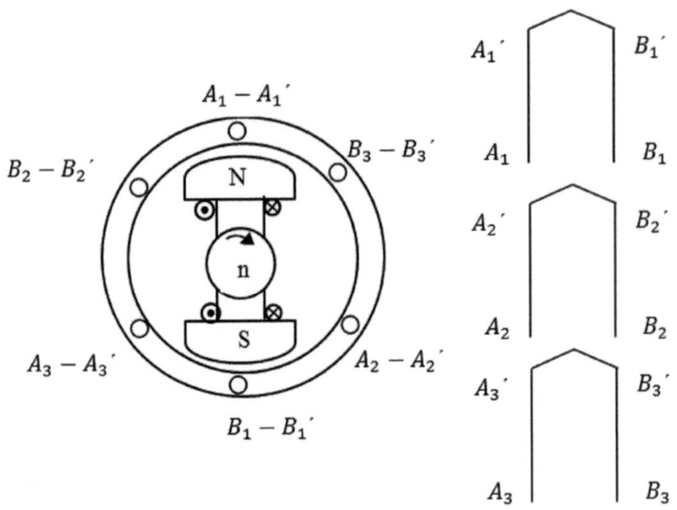

Fig. 6.39 Three-phase two-pole machine (p = 1) winding. *Source* Own elaboration

$$360 \text{ geometric degrees} = p\,360 \text{ electrical degrees} \tag{6.26}$$

Since 360° geometrically always corresponds to (p 360°) electrically, the separation between the phase principles in this case must be 60° geometrically. One can infer 120° electrical of separation from this 60° geometrical separation. Same considerations that apply to stator windings also apply to rotor windings.

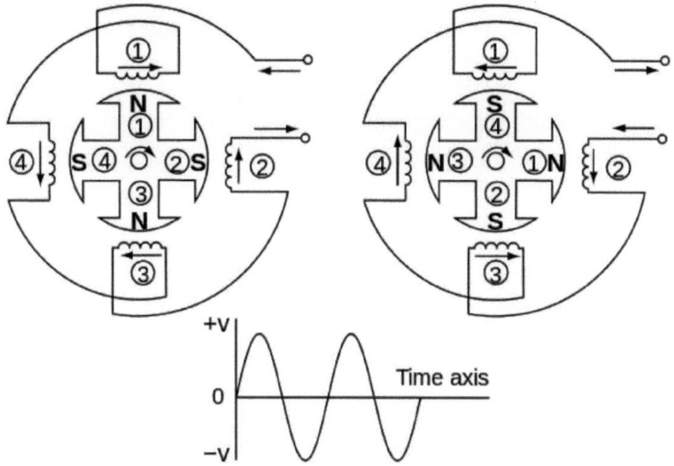

Fig. 6.40 A single-phase, four-pole machine (p = 2). *Source* From wikicommons under free to use license [32]

6.11.2 Winding Factor

Question How do we quantify the impact of these windings' design features on the electrical machine's behavior?

The winding factor (Kb) represents what is the influence of the windings design characteristics on the behavior of the electrical machine. The winding factor is the product of the distribution factor (Kd), the pitch factor (Kp) and the slot slope factor (Ki). We express the winding factor as:

$$K_b = K_d \cdot K_p \cdot K_i \tag{6.27}$$

where Kd is the distribution factor, Kp is the pitch factor, and K_i is the slot tilt factor.

6.11.3 Distribution Factor

In all the examples that we have reviewed the number of slots per pole and phase (K_{pq}) was 1. In practice, however, it is common to design the windings with a distribution factor value K_{pq} greater than 1. This is because, on the one hand, the volume that the active sides of the windings occupy makes it impossible to fit them all together in the same slot. We express the number of slots per pole and phase as:

$$K_{pq} = \frac{\text{Total number of slots}}{\text{number of poles} \cdot \text{number of phases}} = \frac{N_{Slots}}{(2 \cdot p) \cdot q} = 1 \tag{6.28}$$

On the other hand, the distribution of loops over the entire perimeter ($K_{pq} > 1$) allows a better distribution of losses inside the machine. As a result, the machine operates at a uniform temperature and prevents mechanical stress. The distribution factor represents the influence of loops that are not concentrated in a single point, but are distributed over the entire circumference.

We express the distribution factor as:

$$K_d = \frac{sen(90/q)}{K_{pd} \cdot sen\left(\frac{90}{q \cdot K_{pq}}\right)} \tag{6.29}$$

where Q is the number of phases; and K_{pq} is the number of slots per pole and phase.

A simple mathematical analysis shows that:

Assuming a concentrated winding, $K_{pq} = 1$, $K_d = 1$ too. On the other hand, assuming a distributed winding, $K_{pq} > 1$, $K_d < 1$ too.

6.11.4 Pitch Factor

We know that pitch factor appears when the coil pitch differs from the polar pitch. This means that the angular distance between the coil's two active sides may be different from the distance between two successive poles.

We use the following expression to find it:

$$K_p = sen\left(90\frac{Y_B}{Y_p}\right) \tag{6.30}$$

where Y_B is the coil pitch, and Y_p is the polar pitch.

Depending on the ratio of coil pitch to polar pitch, the pitch factor ranges between the values within the following limits:

1. Diametral pitch winding $\rightarrow Y_B = Y_P \rightarrow K_P = 1$
2. Shortened pitch winding $\rightarrow Y_B < Y_P \rightarrow K_P < 1$
3. Long pitch winding $\rightarrow Y_B > Y_P \rightarrow K_P > 1$.

6.11.5 Slot Tilt Factor

The factor for slot inclination arises from the potential inclination of the slots regarding the cylinder generatrix that outlines the surface of the cylinder where they are located.

Its expression is:

$$K_i = \frac{sen\left(\frac{c}{Y_p} \cdot \frac{\pi}{2}\right)}{\frac{c}{Y_p} \cdot \frac{\pi}{2}} \tag{6.31}$$

where C is the slot inclination, measured in slots; and Y_p is the polar pitch, measured in slots.

The analysis of this expression leads to:

Straight slots $\rightarrow c = 0 \rightarrow K_i = 1$.

Sloping slots $\rightarrow c \neq 0 \rightarrow K_i < 1$.

Remember that:

$$\rightarrow \lim_{x \to 0} \frac{\sin x}{x} = \frac{\sin(0)}{0} = \frac{0}{0}$$

According to L'Hopital:

$$= \lim_{x \to C} \frac{f'(x)}{g'(x)} = \lim_{x \to C} \frac{\sin x}{x} = 1$$

Other relationships are:

$$\rightarrow \lim_{x \to C} \frac{1 - \cos x}{x} = 0;$$

$$\rightarrow \lim_{x \to \infty} (1 + \frac{1}{x})^x = e;$$

$$\rightarrow \lim_{x \to C} \frac{e^{kx} - 1}{x} = 1;$$

$$\rightarrow \lim_{x \to C} \frac{a^{kx} - 1}{x} = klna;$$

The analysis of this expression leads to:

If inclined slots $\rightarrow c \neq 0 \rightarrow K_i < 1$.

By definition, none of the factors can surpass unity. Thus, the winding factor, which results from their product, is always bound to be equal or less than one in every constructive scenario.

6.12 Values of Sinusoidal Alternating Current

6.12.1 Basic Generator

After an extensive review of the principles governing electrical machines, let us focus now on find out the magnitude and direction of the emf that the motion of the coil within a magnetic field generates. A voltage waveform appears when the coil rotates inside a magnetic field (Fig. 6.41). Both define the magnitude and direction of the voltage.

It is interesting illustrate a constructive detail of an elemental generator (Fig. 6.42). The coil rotates with angular speed *wr* inside a magnetic field, B, and then induces an electric field E.

For analysis purposes, we shall divide the rotation into some stages (Fig. 6.44). At the starting point, where the angle is 0 degrees, the value of "e" is 0 as well. This is because the poles do not intersect any flux lines. The number of flux lines intersecting increases during the second phase, which causes the value of "e" to increase. In this period, the graph exhibits a positive sinusoidal pattern. During the third phase, the wire reaches an angle close to 90°. More flux lines intersect in this area per unit of time. Consequently, the emf "e" is at its peak here. During phase 4, the conductor shifts away from the maximum point. "E" decreases in the same way it had previously increased. In phase 5, "e" decreases until it again cancels when the conductor passes through 180°. Consequently, "e" increases comparably, but with inverse polarity.

Question Does the graph below Fig. 6.43 correspond to a single or three-phase generator?

The graph (Fig. 6.43) corresponds to an alternating current, as it changes polarity every 180º.

And it is also a single-phase generator because there is only one waveform.

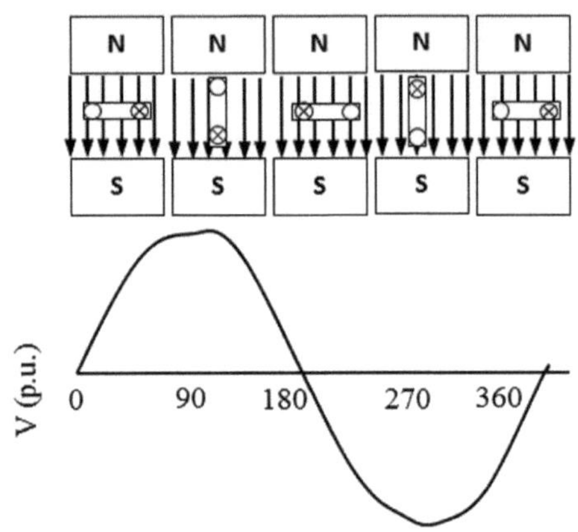

Fig. 6.41 Emf of a coil rotating in a magnetic field. *Source* Own elaboration

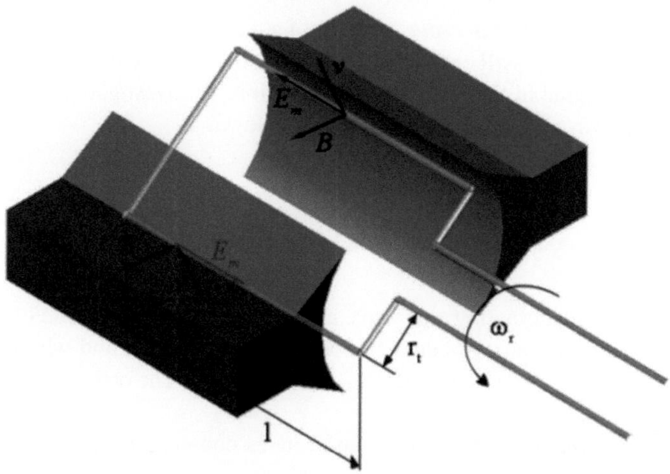

Fig. 6.42 Elemental generator. *Source* From wikicommons under free to use license [33]

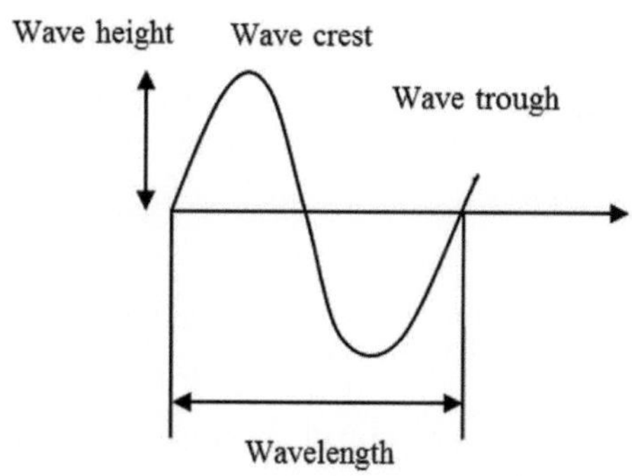

Fig. 6.43 Wave values. *Original Source* Own elaboration

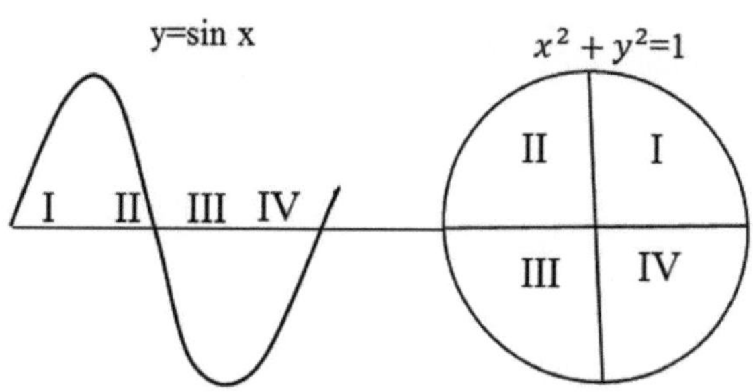

Fig. 6.44 Wave values-cartesian. *Source* Own elaboration

As we see, the instantaneous value of emf between the loop's terminals is a sine function that we express as:

$$e = E_0 \cdot sen(\omega t + \varphi_0) \tag{6.32}$$

Now, we plot the Cartesian representation of a wave (Fig. 6.44). The abscissa axis shows the angles ωt (we will explain ω and t in the following sections). The ordinate axis displays the value of the instantaneous function. To investigate the values of an alternating sine wave, we must first generate a sine wave.

6.12.2 Values Associated with a Sine Wave

We can also plot an alternating sine wave and the projection of the end of a rotating vector whose angular velocity equals the wave's pulsation, ω, and whose modulus equals the maximum value, or amplitude, A_0 (Fig. 6.45). The ordinate axis displays the representation.

Vector $\vec{A}$ has a fixed origin point O and rotates counterclockwise with constant velocity ω. We call this rotating vector, the phasor.

The mathematical function for representing this wave is:

$$a = A_0 \cdot sen(\omega t + \varphi_0) \tag{6.33}$$

To gain a deeper comprehension of the numerical values of a wave from a mathematical and electrical point of view, we analyze the emf that the basic generator generates.

> **Important** Peak value or amplitude.

The highest point or magnitude of an emf signal is the maximum value it reaches during each half-period, which is of the opposite sign in both half-periods, E_0.

> **Important** Peak-to-peak value.

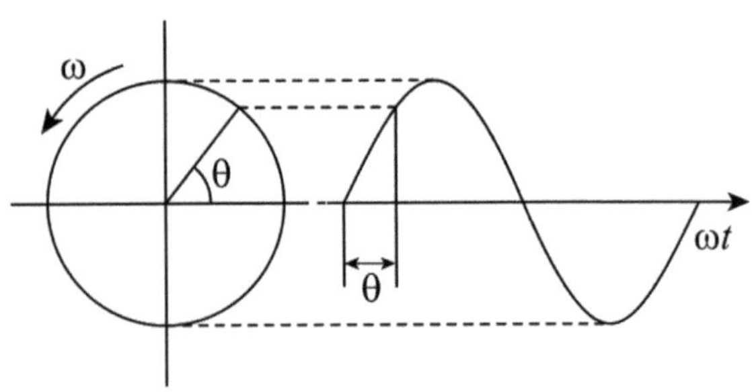

Fig. 6.45 Wave values projection. *Source* Own elaboration

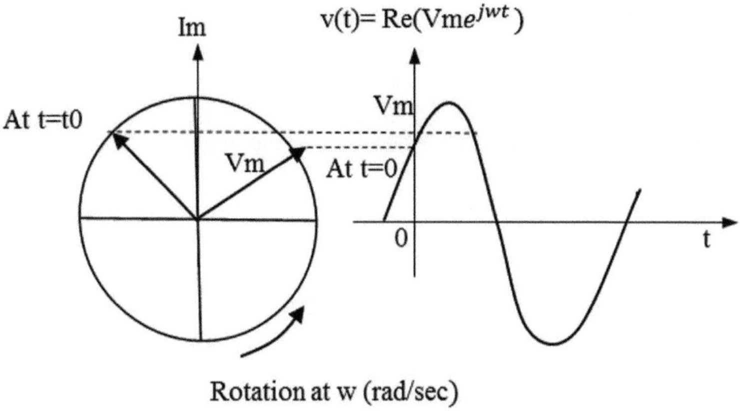

Fig. 6.46 Representation of voltage phasor. *Source* Own elaboration

The peak-to-peak value is twice the maximum value as defined by convention. Assuming this is an electromotive wave, we express the emf peak-to-peak value as:

$$E_{pp} = 2 \cdot E_0 \tag{6.34}$$

Important Pulse.

The pulsation ω is the angular velocity at which the generator rotates. As we know we measure it in rad/s. The speed and the phasor speed are the same. We can plot the phasor V e^{jwt} (Fig. 6.46). The figure on the left shows a sine wave rotating counterclockwise. The figure on the right shows its projection on the real axis, as a function of time.

We define the angular speed as:

$$\omega = \frac{\varphi}{t} \tag{6.35}$$

φ is the angle that this wave travels in time, t.

Important Period.

It is the time in seconds it takes for a loop to complete one revolution. Observe that a wave travels across an angle $\varphi = 2\pi \,\text{rad}\left(360^{\circ}\right)$ in a time t that equals the period T. We also define it as the time it takes to complete one cycle. From (Eq. 6.35), we arrange the expression to find the angle that travels by the loop or by the phasor as follows:

$$\varphi = \omega t(6.36) \tag{6.36}$$

This equation demonstrates that when t = T, the angle the wave travels is 2π radians. We can therefore express it as:

$$\omega t = 2\pi \tag{6.37}$$

Already clearing T, we arrive at the following expression:

$$T = \frac{2\pi}{\omega} \tag{6.38}$$

Important Frequency.

Frequency measures the number of cycles the wave travels in one second. We determine it by calculating the number of waves, periods or cycles that completes within this time frame. While the above definition is correct, it cannot address what frequency means in practical terms. In other words, frequency corresponds to the number of cycles the loop makes per second.

Frequency is the reciprocal of the period and we measure it in hertz (Hz) within the international system.

Question What is the unit of frequency? The hertz, or cycles per second is the fundamental physical unit of frequency. We express it as:

$$f = \frac{1}{T} = \frac{w}{2\pi}\text{Hz}\left(\text{cycless}^{-1}\right) \tag{6.39}$$

To answer the question, one Hertz equals one cycle per second, with the second being one of the seven fundamental physical units. The unit of frequency is named after Heinrich Hertz, one of the first German researchers of radio wave transmissions. Note that Heinrich Hertz said: "I do not think that the radio waves I have discovered will have any practical application".

Question What are the frequency values of the alternating currents used in Europe and the USA, and what are their respective pulsations?

The frequency value of the most common electric used alternating currents systems is 50 Hz within Europe and 60 Hz within the USA. The corresponding pulsations are $\omega = 100\pi$ and 120π rad/s. Remember that $\text{f} = \frac{1}{\text{T}} = \frac{\omega}{2\pi}$ Hz (cicles,·s^{-1}). As a result, for the European case, the alternating current changes direction 100 times per second.

Important Phase angle.

The phase angle in a sine wave is the angle ($\omega t + \varphi$) that appears when we compare with the initial phase (t = 0). In phenomena wherein the periodic magnitude fluctuates with respect to time, the term phase pertains to the specific point, location, or condition in which the phenomenon is being analyzed. We define the phase as the period that has elapsed since the function passed through the reference position, which is termed as the "phase origin".

6.12.2.1 Two-Phase System

It is a system comprising two voltages of similar magnitude and frequency (Fig. 6.47). This is a two-phase system, but differing in phase by 90°.

When we have two sine functions in an electrical machine, the phase difference determines the position between them. We call this distance the phase difference.

Important Instantaneous value.

The instantaneous value of the emf (e) is the value it holds at any instant (Fig. 6.48).

If we multiply E_0 by the sine of the angle ($\omega t + \varphi 0$), we are projecting the phasor on the ordinate axis (Eq. 6.40). Then, we define the instantaneous emf value.

$$e = E_0 \cdot sen(\omega t + \varphi_0) \tag{6.40}$$

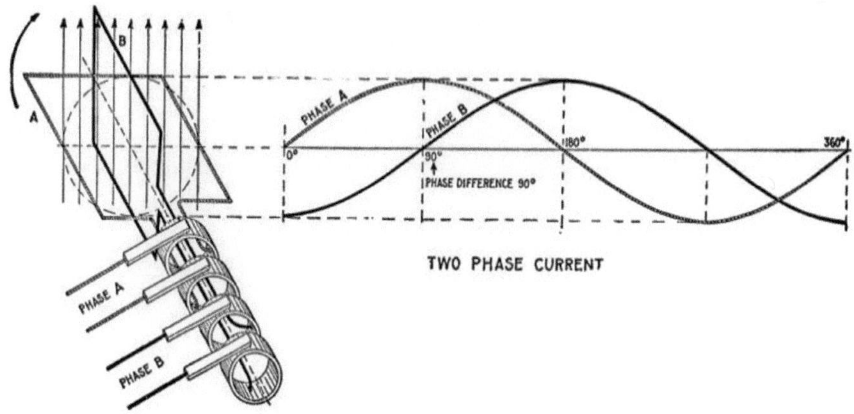

Fig. 6.47 Two Phase system. *Source* From wikicommons under free to use license [34]

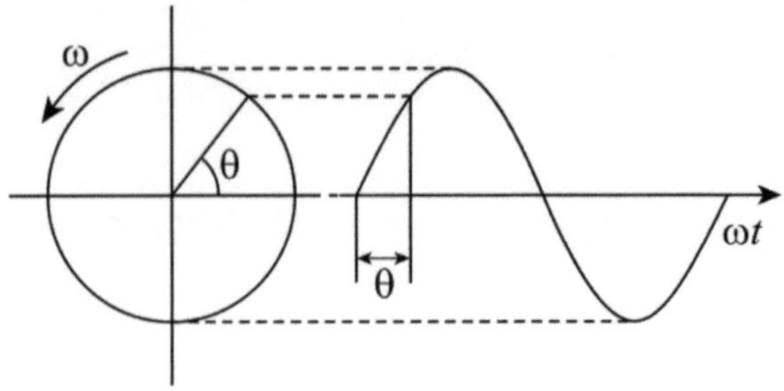

Fig. 6.48 Instantaneous value of a sinusoidal waveform. *Source* Own elaboration

Important Average value over a half period.

The mean value is the value that moving coil type devices measure (Fig. 6.49a). In addition, we also show its symbol (Fig. 6.49b).

The needles' deflections correlate to the currents flowing through the moving coil. Hence, the interaction between electromagnetic effects of the magnet field and the currents passing through the voice coil determines the needle's position. The balance position is the average value of the needle's positions. We define the mean as the arithmetic average of the instantaneous values over a half-period. However, the average value of a sinusoidal function during a period is zero. Consequently, we only need to examine a single half-period.

An indicator of the average value of the needle positions is the equilibrium position (Fig. 6.50).

Question Can we obtain the average value of a half period from the maximum value?

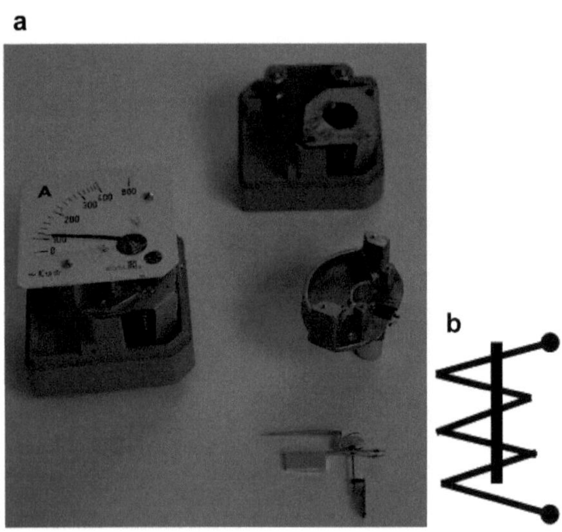

Fig. 6.49 **a** and **b** Moving coil ammeter and moving Iron Instrument Symbol. *Source* From wikicommons under free to use license [35]

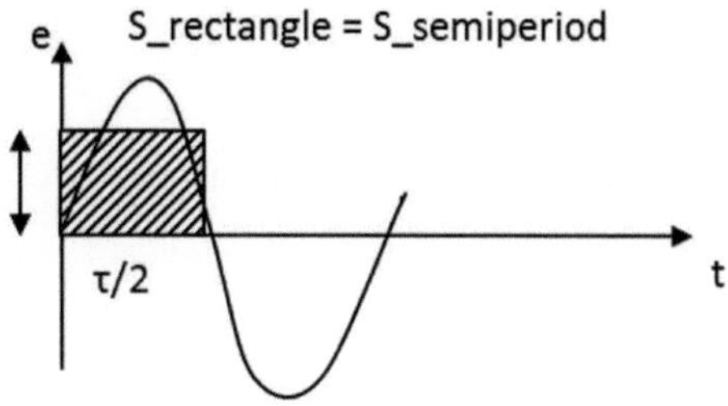

Fig. 6.50 Mean value in a sinusoidal waveform. *Source* Own elaboration

The conjecture we have examined hitherto on the average value of a half period is still purely theoretical. Irrespective of the engineering specialty, it's essential to comprehend the problem in order to find a viable solution. It is, after all, an essential objective of an engineer to solve complex problems. At the close of the 1800s, Lord Kelvin was renowned among his pupils at the University of Glasgow for imparting the maxim that "what cannot be defined cannot be measured. What is not measured cannot be improved. And what is not improved is always degraded". It bears reminding, though, that what cannot be demonstrated holds no utility for learning.

There is little value in blindly memorizing a set of exercises and theory when the equations that we use to solve these exercises are seemingly arbitrary. As such, we shall demonstrate the step-by-step derivation of the equation for finding the mean value in a half-period, and the rms value, both frequently employed in electrical engineering.

We will therefore begin by illustrating the mean value of half a period.

Then we have (Fig. 6.50)

$$\mathrm{A_{rectangle}} = \mathrm{A_{semiperiod}}$$

In this case, the area of the rectangle is:

$$A_{\text{rectangle}} = E_m \cdot \frac{T}{2} \tag{6.41}$$

We determine the area of the half-period as:

$$A_{\text{semiperiod}} = \int_0^{T/2} e \cdot dt \tag{6.42}$$

Equating expressions (6.41) and (6.42), we have:

$$E_m \cdot \frac{T}{2} = \int_0^{T/2} e \cdot dt \tag{6.43}$$

And by clearing $\mathrm{E_m}$:

$$E_m = \frac{1}{T/2} \int_0^{T/2} e \cdot dt \tag{6.44}$$

Substituting the instantaneous value "e" for its expression, we have Eq. 6.45:

$$E_m = \frac{2}{T} \int_0^{\frac{T}{2}} E_0 \cdot sen\omega t \cdot dt = \frac{2E_0}{T} \int_0^{\frac{T}{2}} sen\omega t \cdot dt = \frac{2E_0}{T} \cdot [-\cos \omega t]_0^{\frac{T}{2}} =$$

$$\frac{2E_0}{T\omega} \cdot [-((-1) - (+1))] =$$

$$\frac{2 \cdot 2 \cdot E_0}{T\omega} \cdot [-((-1) - (+1))] =$$

$$\frac{2 \cdot 2 \cdot E_0}{T\omega} =$$

Recalling that: $T = \frac{2\pi}{\omega}$ and clearing w we have: $\omega = \frac{2\pi}{T}$.
As a consequence, E_m is:

$$E_m = \frac{4 \cdot E_0}{\frac{2\pi}{T} \cdot T} = \frac{2}{\pi} \cdot E_0 \tag{6.45}$$

If we had considered current wave instead of electromotive forces, we would have come to the same conclusion. Then, we have:

$$I_m = \frac{2}{\pi} \cdot I_0 \tag{6.46}$$

In the same way we express the average voltage U_m, as:

$$U_m = \frac{2}{\pi} \cdot U_0 \tag{6.47}$$

Important Root Mean Square (RMS) value.

In electrical circuit calculus we often use rms values of alternating voltages and currents because electrodynamic devices measure quadratic quantities.

Regardless of the sign, measurements are proportionate to the square of the quantity they are measuring. The rms value of alternating current is equivalent to that of direct current, which generates the same amount of heat as the alternating current when apply it to a pure ohmic resistance.

This definition is valid for electromotive force, voltage and power. The literature uses the term rms for rms voltage or rms power to refer to the rms characteristic. For example, 10 VRMS or 15 WRMS mean 10 RMS volts or 15 RMS watts respectively. The rms value of a variable quantity is the square root of the mean of the instantaneous values over time. For the electromotive force, we express it as:

$$E = \sqrt{\frac{1}{T}\int_0^T e^2 dt} \tag{6.48}$$

The rms value is:

$$V_{RMS} = \frac{V_{PEAK}}{\sqrt{2}} \approx V_{PEAK} \times 0,707 \tag{6.49}$$

Question How can we prove that the root mean square value is indeed equal to the maximum value divided by the square root of two?

This is something that we are going to verify now on. From definition we know that:

$$Q_{dc} = Q_{ac} \tag{6.50}$$

Q_{dc} is the energy that a pure ohmic resistor turns into heat when direct current passes through it. Q_{ca}, on the other hand, is the energy that the same resistor turns into heat when alternating current passes through it.

Heating a resistor

Because $P = I^2R$ and $P = V^2/R$, we express the total energy which converts into thermal energy in different manners.

Thermal energy: $E = P \cdot t; E = I^2R \cdot t; E = \left(V^2/R\right) \cdot t.$

If condition of Eq. 6.50 holds true and the heat under DC or AC current and the resistance is the same, then it is true that:

$$U_{dc} = U_{ac} \tag{6.51}$$

With U_{dc} representing the voltage in the DC resistor and Uac representing the rms value of the AC voltage, both the rms and direct current (DC) values express the same for the voltage (U), emf (E), current (I) and power (P). It also represents the thermal energy that the direct current dissipates in a resistor R.

$$Q_{dc} = R \cdot I^2 \cdot t \tag{6.52}$$

For a time (t), equal to the period T, we can express Eq. 6.52 as:

$$Q_{dc} = R \cdot I^2 \cdot t \tag{6.53}$$

and

$$dQ_{ac} = R \cdot I^2 dt \tag{6.54}$$

For a period (T), Eq. 6.54 becomes:

$$Q_{ac} = R \cdot \int_0^T I^2 \cdot dt \tag{6.55}$$

Equating (6.53) and (6.54), we obtain:

$$I^2 \cdot T = \int_0^T i^2 \cdot dt \tag{6.56}$$

And clearing I^2:

$$I^2 = \frac{1}{T} \int_0^T i^2 \cdot dt \tag{6.57}$$

Extracting the square root of the Eq. 6.57 results:

$$I = \sqrt{\frac{1}{T} \int_0^T i^2 \cdot dt} \tag{6.58}$$

This equation defines the average value.
If we recall the value of current i we have:

$$I^2 = \frac{1}{T} \int_0^T I_0^2 \cdot sen^2(\omega t) \cdot dt = \frac{I_0^2}{T} \int_0^T sen^2(\omega t) \cdot dt \tag{6.59}$$

Let us now review the basic trigonometric identity and the double angle identity. To solve the integral (Eq. 6.59), it is important to take into account the following:

$$\begin{cases} \text{sen}^2(\omega t) + \cos^2(\omega t) = 1 \\ \text{sen}^2(\omega t) - \cos^2(\omega t) = -cos(2\omega t) \end{cases}$$

Adding both equations:

$$2 \cdot sen^2(\omega t) = 1 - cos(2\omega t)$$

Then, we have:

$$sen^2(\omega t) = \frac{1 - cos(2\omega t)}{2}$$

Substituting this result in Eq. 6.59 yields:

$$I^2 = \frac{I_0^2}{2 \cdot T} \cdot \int_0^T dt - \frac{I_0^2}{2 \cdot T} \cdot \int_0^T \cos(2\omega t) \cdot dt \tag{6.60}$$

Next, we solve the integral of cos 2 × dx that results 0.5 (sin 2x). Then:

$$\int \cos 2x dx$$

Let u = 2x,

$$\frac{du}{dx} = 2$$
$$\frac{1}{2} du = dx$$

$$\int \cos 2x dx = \int \cos u \frac{1}{2} du$$

$$\int \cos u \frac{1}{2} du = \frac{1}{2} sinu = \frac{1}{2} sin2x$$

Substituting in Eq. 6.60 yields:

$$I^2 = \frac{I_0^2}{2 \cdot T} \cdot [t]_0^T - \frac{I_0^2}{2 \cdot T \cdot 2 \cdot \omega} \cdot [sin(2\omega t)]_0^T \tag{6.61}$$

Recalling that $\sin(2\omega t) = \sin(4\pi) = 0$, we obtain:

$$I^2 = \frac{I_0^2}{2 \cdot T} \cdot T - \frac{I_0^2}{4 \cdot \omega \cdot T} \cdot 0 \rightarrow I^2 = \frac{I_0^2}{2} \rightarrow I = \frac{I_0}{\sqrt{2}} \tag{6.62}$$

Recall that the pulsation, ω, equals 2πf or, equivalently, 2π/t. Consequently, we may express the root mean square value of the voltage as:

$$I = \frac{I_0}{\sqrt{2}} \tag{6.63}$$

If we analyze electromotive force or voltage instead of current, we obtain the same result.

$$U = \frac{U_0}{\sqrt{2}} \tag{6.64}$$

$$E = \frac{E_0}{\sqrt{2}} \tag{6.65}$$

After obtaining the same equation at the beginning of this section, we demonstrate that the shape factor of a sine wave is the quotient between the rms value and the mean value. Therefore, its value is expressed as:

$$K_F = \frac{E}{E_m} = \frac{\frac{E_0}{\sqrt{2}}}{\frac{2}{\pi} \cdot E_0} = \frac{\pi}{2 \cdot \sqrt{2}} = 1,11 \tag{6.66}$$

Important Form factor.

The form factor of a sine wave is the ratio of the rms value and the arithmetic mean value. It is important to note that this calculation only applies to sine waves. Therefore, its value is:

$$K_F = \frac{E}{E_m} = \frac{\frac{E_0}{\sqrt{2}}}{\frac{2}{\pi} \cdot E_0} = \frac{\pi}{2 \cdot \sqrt{2}} = 1,11 \tag{6.67}$$

This factor varies depending on the signal's waveform and differs for signals with varying shapes. Please refer to Table 6.2 for more information on this factor.

For instance, for a sine waveform, this coefficient is 1.11. For other signal types, however, the value may vary depending on signal deformation.

Important Peak factor.

The crest or amplitude factor represents the ratio of the maximum value of voltage and the rms value of the wave. This applies for sinusoidal signals:

$$K_a = \frac{E_0}{E} = \frac{E_0}{\frac{E_0}{\sqrt{2}}} = \sqrt{2} \tag{6.68}$$

Table. 6.2 Associated values in a waveform

Waveform	RMS	Crest factor	Form factor
Sine wave	$\frac{1}{\sqrt{2}}$ or 0.7071	$\sqrt{2}$ or 1.414	1.11
Triangular wave	$\frac{1}{\sqrt{3}}$ or 0.577	$\sqrt{3}$ or 1.732	1.154
Square wave	1	1	1
Saw tooth wave	$\frac{1}{\sqrt{2}}$ or 0.577	$\sqrt{3}$ or 1.732	1.154

Source Own elaboration

The peak value of a waveform is waveform-dependent. For a sinusoidal waveform, the peak value is 1.41 times greater than that of the rms value. However, for distorted signals, the peak value may significantly differ from the rms value of a sinusoidal signal. Thus, for a sinusoidal signal, the peak value is 21.2 A when an rms value of 15 A is at a particular point. For a distorted signal, such as a pure triangular signal, the peak value could reach up to 26A.

Example 6.5 Consider a multimeter that is configured to gauge AC voltage, we assess the voltage present at a socket within the classroom. The operating grid is 230 V at a frequency of 50 Hz.

Determine:

(a) The maximum value, the mean value, the amplitude factor, the period and the pulsation.
(b) Draw the u-shaped curve for f (t) over one complete cycle.
(c) The instantaneous voltage value 0.003 s after passing through the point $\frac{U_0}{2}$.

Answer.

If the rms value is $U = 230$ V, the maximum value will be: $U_0 = \sqrt{2} \cdot 230 = 325,27$ V.

Applying (Eq. 6.46), the average value is: $U_m = \frac{2}{\pi} \cdot U_0 = \frac{2}{\pi} \cdot 325,27 = 207,07$ V. On the contrary, the amplitude factor, K_a is:

$$K_a = \frac{U_0}{U} = \frac{325,27}{230} = \sqrt{2}$$

The period T represents the time to the function to finish one cycle. Therefore, when the frequency equates to f = 50 Hz = 50 cycles-s-1, T will be:

Next, we obtain the pulsation ω in rad/s:

$$\omega = 2 \cdot \pi \cdot f = 100\pi = 314 \text{ rad} \cdot s^{-1}$$

Figure 6.51 illustrates the function and shows the instantaneous variation in voltage. The x-axis plots a complete cycle, with time in seconds and angles in radians, bearing in mind that 1 cycle is equal to 2π radians and lasts 0.02 s.

Because it is a sinusoidal wave, we express it as:

$$u = U_0\, sen(wt)$$

Substituting values we have:

$$u = 325,27\, sen\,(314,16t)(V)$$

The angle *wt* corresponds to an instantaneous voltage of:

Fig. 6.51 Example 6.5.
Source own elaboration

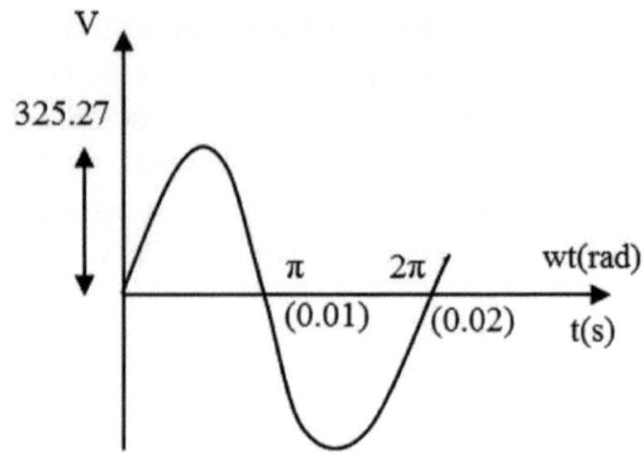

$$U = U_0/2$$

$$U_0/2 = U_0 sen\,(wt)$$

In other words:

$$sen,\ wt = \frac{1}{2};\ \text{Therefore: } wt = 30^\circ = \frac{\pi}{6}\text{rad}$$

On the contrary, 0.003 s corresponds to an angle of:

$$wt = 314,16 \times 0,003 = 0,942\,\text{rad}$$

The value therefore corresponds to an angle covered by the voltage wave of $(0.942 + \pi/6)$ radians. This is the value of the voltage at that particular time:

$$u = U_0\,sen\,(wt) = 325,27\,sen\left(0,942 + \frac{\pi}{6}\right) = 323,47(V)$$

6.12.3 Generating from Renewable Energy Sources: The Sun

It is worth mentioning others ways to generate electricity that are widely used today in most energy systems. Despite the fact that the energy that from the sun we collect is not considered an electrical machine is extensively used for electricity generation, among others. There are several technologies that we use to collect the energy from the sun and generate electricity. Depending on the solar panel used electrical engineers model its behavior as from the experiments they perform. According to the results they gather different models are used to perform the equivalent circuit of a solar cell (Fig. 6.52). In addition, as we have previously seen we attach a generator

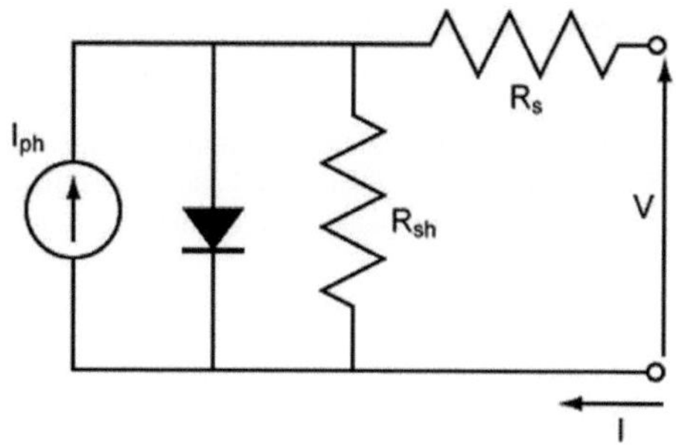

Fig. 6.52 Equivalent circuit of a solar cell with symbols. *Source* From wikicommons under free to use license [36]

to a water turbine. Similarly, we can attach it to a wind turbine and generate electricity from the wind energy. Independently the source of energy, once we deliver the current through the wires we model it according to laws we have presented as well others.

6.13 Summary

This chapter establishes the fundamental concepts behind electrical machines. It begins by classifying machines into two main categories: static machines (like transformers) and rotating machines (like motors and generators). For static machines, it details different types such as autotransformers and instrument transformers, and describes their core active elements: the primary and secondary electrical circuits and the magnetic core.

The chapter then shifts to rotating machines, outlining their constituent parts and explaining the core principle of energy conversion, emphasizing the essential role of a magnetic field. It delves into the physics of operation, covering the generation of magnetic fields, key magnetic quantities, and fundamental laws, including the effects of both direct and alternating current excitation. Finally, the chapter applies these principles to the electrical circuits of machines, discussing concepts like the winding factor and concluding with the analysis of sinusoidal alternating current systems and a basic generator model.

References

1. E. at G. Wikipedia, Deutsch: Bild von Schnittbandkerntrafo (2008). Accessed 23 Dec 2023. [Online]. https://commons.wikimedia.org/wiki/File:Schnittbandkerntrafo-07kva.JPG
2. G. Shape svg by, Schemat ogólny autotransformatora (2008). Accessed 10 Dec 2023. [Online]. https://commons.wikimedia.org/wiki/File:Autotransformator_schemat.svg
3. C. by I.-J.C. Photographer, English: Low Voltage Ring Type & Plastic Case Current Transformers (2008). Accessed 10 Dec 2023. [Online]. https://commons.wikimedia.org/wiki/File:ITL-Family.jpg

4. Wordtwist, English: instrument potential transformer showing nameplate, instantaneous polarity markings in two styles (dot and X1) 120:120V; 600Volt class; 0.3W, 0.6X (0.3% and 0.6%) accuracy; isolation for phase shifting potentials (2013). Accessed 10 Dec 2023. [Online]. https://commons.wikimedia.org/wiki/File:Instrument_Transformer.jpg
5. Constant314, English: A 1:1 isolation transformer with extra insulation b etween primary and secondary, usually constructed from a few layers of mylar tape. Accessed 10 Dec 2023. [Online]. https://commons.wikimedia.org/wiki/File:Isolation_Transformer_Simple_with_Dielectric_Barrier.jpg
6. A. Pedreira, Toroidal transformer (2006). Accessed 23 Dec 2023. [Online]. https://commons.wikimedia.org/wiki/File:TRFtoro.PNG
7. Daderot, English: Exhibit in the stationary steam engine collection—New England Wireless & Steam Museum—East Greenwich, Rhode Island, USA (2017). Accessed 20 Dec 2023. [Online]. https://commons.wikimedia.org/wiki/File:Bullock_Electric_Co._generator,_Cincinnati,_133_amperes,_120_volts_full_load_-_New_England_Wireless_%26_Steam_Museum_-_East_Greenwich,_RI_-_DSC06736.jpg
8. Wikipedia. Water turbine - edit1.svg. Accessed 10 Dec 2023. [Online]. https://commons.wikimedia.org/wiki/File:Water_turbine_-_edit1.svg
9. Wikipedia. Faraday emf experiment.svg. Accessed 10 Dec 2023. [Online]. https://commons.wikimedia.org/wiki/File:Faraday_emf_experiment.svg
10. Wikipedia. Innenpolmaschine.svg. Accessed 23 Dec 2023. [Online]. https://commons.wikimedia.org/wiki/File:Innenpolmaschine.svg
11. L.R. Krol, English: An electric current I flowing through a conductor produces a magnetic field B (2020). Accessed 9 Dec 2023. [Online]. https://commons.wikimedia.org/wiki/File:Magnetic_field_produced_by_current.svg
12. Zureks, English: Schematic representation of Earth's magnetic field linesPolski: Schematyczne przedstawienie linii pola magnetycznego Ziemi (2012). Accessed 9 Dec 2023. [Online]. https://commons.wikimedia.org/wiki/File:Earth%27s_magnetic_field,_schematic.svg
13. Självinduktans. Accessed Dec 2023. [Online]. https://commons.wikimedia.org/wiki/File:Sj%C3%A4lv-induktans.svg
14. Wikipedia. Magnetic field due to dipole moment.svg. Accessed 9 Dec 2023. [Online]. https://commons.wikimedia.org/wiki/File:Magnetic_field_due_to_dipole_moment.svg
15. Letanure, English: electroima (2008). Accessed 9 Dec 2023. [Online]. https://commons.wikimedia.org/wiki/File:Eletroima.png
16. Wikipedia. F. right-hand rule svg (2020). Accessed 9 Dec 2023. [Online]. https://commons.wikimedia.org/wiki/File:Coil_right-hand_rule2.svg
17. Wikipedia. GrammeMachine.jpg. Accessed 9 Dec 2023. [Online]. https://commons.wikimedia.org/wiki/File:GrammeMachine.jpg
18. Wikipedia. P. at F. Wikipedia, Français: Schéma de la carcasse d'un transformateur monophasé bas de gamme. Dessin personnel réalisé pour l'article transformateur électrique de yohann de olivera (2005). Accessed 24 Dec 2023. [Online]. https://commons.wikimedia.org/wiki/File:Carcasse_transformateur_monophas%C3%A9.png
19. Backsideficker, English: example magnetic circuit to demonstrate voltage divider. Accessed 9 Dec 2023. [Online]. https://commons.wikimedia.org/wiki/File:Spannungsteiler_Bsp3.svg
20. Zureks, English: Magnetic permeability (not to scale): B—flux density, H—magnetic field, μf—permeability of ferromagnets (and ferrimagnets), μp—permeability of paramagnets, μ0—permeability of free space (4*π*10–7 H/m), μd—permeability of diamagnets (2009) Accessed 9 Dec 2023. [Online]. https://commons.wikimedia.org/wiki/File:Permeability_by_Zureks.svg
21. Zureks, Polski: Prawo przepływu (prawo Ampera) N*I = H1*l1 + H2*l2 (2009). Accessed 24 Dec 2023. [Online]. https://commons.wikimedia.org/wiki/File:Prawo_przeplywu.svg
22. Wikipedia. Hysteresiskurve.svg—Wikipedia, la enciclopedia libre». Accessed 9 Dec 2023. [Online]. https://commons.wikimedia.org/wiki/File:Hysteresiskurve.svg
23. Wikipedia. Magnetization curves.svg. Accessed 24 Dec 2023. [Online]. https://commons.wikimedia.org/wiki/File:Magnetization_curves.svg

24. Zureks, English: Magnetic permeability of ferromagnetic (and ferrimagnetic) materials (2009). Accessed 9 Dec 2023. [Online]. https://commons.wikimedia.org/wiki/File:Permeability_of_ferromagnet_by_Zureks.svg
25. Д. И. vectorization, Deutsch: Verhältnisse an der magnetischen Hysteresekurve bei einem Rindgkernspeicher (2020). Accessed 9 Dec 2023. [Online]. https://commons.wikimedia.org/wiki/File:Ringkernspeicher_hysteresekurven.agr.svg
26. Wikipedia. Chetvorno: A Barlow wheel, an early experiment demonstrating the principle of a homopolar electric motor, invented by English mathematician Peter Barlow in 1822, on display at the SPARK Museum of Electrical Invention, Bellingham, Washington, USA. It consists of a star shaped wheel, pivoted so it can turn, with the points dipping into a trough of the liquid metal mercury in the wooden base. During operation a horseshoe magnet is placed on the wooden base, with its poles embracing the wheel. When an electric current is passed between the hub of the wheel and a contact in the mercury, the wheel rotates (2018). Accessed 8 Dec 2023. [Online]. https://commons.wikimedia.org/wiki/File:Barlow_wheel_1848_-_Spark_Museum.jpg
27. Wikipedia. Cross section of a simple isolation transformer with no special features. Color code: black circles = primary winding, red circles = secondary winding, blue circles = electrostatic shield winding, pale yellow = bobbin, bright yellow=dielectric barrier, cross hatched = magnetic core, gray = enclosure. Accessed 24 Dec 2023. [Online]. https://commons.wikimedia.org/wiki/File:Isolation_Transformer_Simple.jpg
28. Wikipedia. Deutsch: Bild von UI-Kern Trafos (2008). Accessed 24 Dec 2023. [Online]. https://commons.wikimedia.org/wiki/File:UI-tr-gr.JPG
29. Vonvikken, English: Switched reluctance motor with 6 stator poles and 4 rotor poles (2009). Accessed 24 Dec 2023. [Online]. https://commons.wikimedia.org/wiki/File:Switched_reluctance_motor_6-4.svg
30. haade, 2 Pole DC motor PM. 2007. Accessed 24 Dec 2023. [Online]. https://commons.wikimedia.org/wiki/File:Motor_DC_2Pole.png
31. Wikipedia. Accessed 24 Dec 2023. [Online]. https://commons.wikimedia.org/wiki/File:Alternador.svg
32. Wikipedia. Diagram of single-phase generator with four poles.svg—Wikipedia». Accessed 6 Dec 2023. [Online]. https://commons.wikimedia.org/wiki/File:Diagram_of_single-phase_generator_with_four_poles.svg
33. N. autore leggibile automaticamente L. zuò presunto, autore (2006). Accessed 8 Dec 2023. [Online]. https://commons.wikimedia.org/wiki/Image:Schema_macchina_elettrica.png?uselang=it
34. Wikipedia. Hawkins Electrical Guide—3phase Elementary 3wire.jpg—Wikimedia Commons». Accessed 8 Dec 2023. [Online]. https://commons.wikimedia.org/wiki/File:Hawkins_Electrical_Guide_-_3phase_Elementary_3wire.jpg
35. Pedalito, Deutsch: Aufbau eines Dreheisenmesswerks (Dreheisenmessgerät) (2021). Accessed 8 Dec 2023. [Online]. https://commons.wikimedia.org/wiki/File:Dreheisenmesswerk_-_ge%C3%B6ffnet_und_auseinandergenomme_beschriftet.png
36. Equivalent circuit of a solar cell (2025). Accessed 22 Sept 2025. [Online]. https://commons.wikimedia.org/wiki/File:SolarCell-EquivalentCircuit3.PNG

Chapter 7
Power Transformers

Abstract Power transformers adjust the voltage between different parts of an electrical system to allow it to operate. The main parts of a transformer are the iron core and the coils, although others components ensure optimal functionality. Copper and iron losses affect the transfer of power between high and low-voltage terminals in a transformer. This chapter revises the equivalent circuit and operating modes of power transformers.

Keywords Power transformers · Electrostatic and electrodynamics · Magnetic circuit and two electrical windings · Electricity applications · Conductors and isolators

7.1 Introduction

Power transformers adjust voltage levels between different parts of the Electric Power System. We need power transformers to increase, and decrease voltage levels (Fig. 7.1).

The transformer comprises a magnetic circuit and two electrical windings, as described above. The different configurations of a transformer include the laminated iron core (Fig. 7.2).

The primary winding is the one through which the power enters the transformer, the side through which it receives the energy. In contrast, the secondary winding is the one to which we connect the load, that is, the side through which it delivers the energy. We find real power transformers in the facilities of the electricity network (Fig. 7.3).

A step-up transformer is a device that has a lower input voltage than its output voltage. In such a transformer, the low voltage (LV) side is the primary, while the high voltage (HV) side is the secondary. In contrast, we classify a transformer as a step-down transformer when the input voltage is greater than the output voltage. In this configuration, the high-voltage side is the primary and the low-voltage side is the secondary.

D. Borge-Diez et al., *Fundamentals of Electrical Power Engineering*, Lecture Notes in Electrical Engineering 1540, https://doi.org/10.1007/978-3-032-18904-2_7

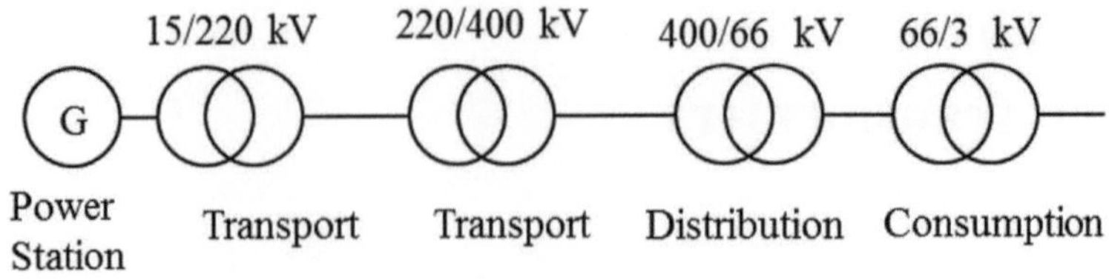

Fig. 7.1 Electrical power system. *Source* Own elaboration

Fig. 7.2 Electrical transformer. *Source* From wikicommons under free to use license [1]

Fig. 7.3 Power transformer. *Original Source* From wikicommons under 3.0 CC use license [2]

7.2 Parts of a Power Transformer

In addition to structural elements, we classify the components of a power transformer in:

Active parts:

This label encompasses the magnetic and electric circuits, which are the elements responsible for the reception and transmission of electrical energy within the transformer. The input/output system enables the interconnection of the transformer's electrical circuit with its input and output power lines. The input and output system of a transformer comprises primarily high-voltage terminals (insulators with smaller conductors) and low-voltage terminals (insulators with larger conductors) (Fig. 7.4). In addition, the transformer comprises a number of other components, including the cooling system, which plays an important role in the transformer's functionality.

Parts in a power transformer include:

1. Tank
2. Lid
3. Conservator tank
4. The oil level indicator (end of conservator tank)
5. Buchholz relay for detecting gas bubbles after an internal fault
6. Piping to conservator tank and Buchholz relay
7. Tap changer to change output voltage

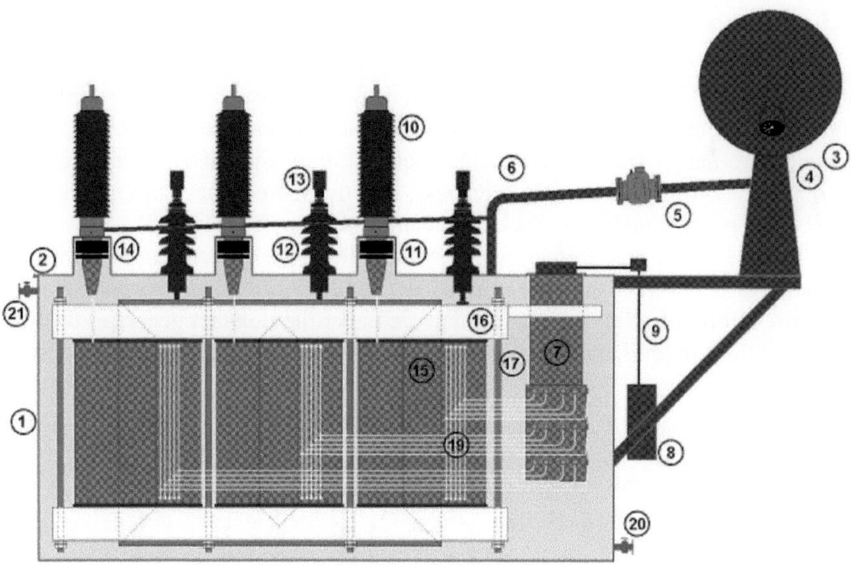

Fig. 7.4 Parts in a power transformer. Original from Fluppe37 at nl.wikipedia. *Source* From wikicommons under 3.0 CC use license [3]

8. The motor drive of the tap changer (can be controlled by an automatic voltage regulator)
9. Drive shaft for tap changer
10. High voltage (HV) bushing connects the internal HV coil with the external HV grid
11. High voltage bushing current transformers for measurement and protection
12. Low voltage (LV) bushing connects LV coil to LV grid
13. Low voltage current transformers
14. Bushing voltage-transformer for metering the current through the passing bushing
15. Core
16. Yoke of the core
17. Limbs connect the yokes and hold them up
18. Coils
19. Internal wiring between coils and tap changer
20. Oil release valve
21. Vacuum valve.

These components are necessary for good transformer operation. As other electrical machines during operation it emits heat, which must dissipate into the surrounding environment in order to maintain the operational temperature within acceptable limits.

7.3 Losses in Power Transformers

The heat sources of a power transformer include transformer losses, hysteresis losses, and eddy current losses. Transformer no-load losses include hysteresis losses, which depend on re-magnetization factors such as quality, frequency, and flux density. Eddy current losses, which depend on thickness, resistivity, frequency, and flux density, also contribute to transformer losses and can be represented (Fig. 7.5).

When the power transformer operates under load, it experiences several losses, including the i^2r loss and the eddy current loss. The eddy current loss is a phenomenon that arises from metallic components that are under the flux that generates the load current in the windings and leads. The evacuation process entails circulating coolant

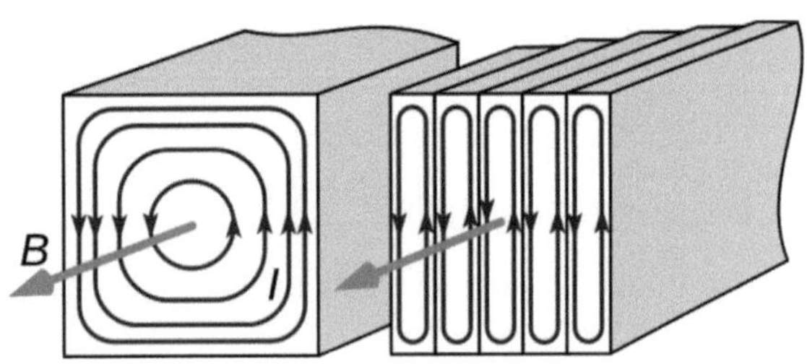

Fig. 7.5 Eddy current losses. *Source* From wikicommons under free to use license [4]

(mineral oil) over the surface of the tank or vat, with some instances cooling fins and/or fans to enhance air circulation and facilitate heat dissipation.

7.3.1 Protection Systems

Another important component of a transformer is the Buchholz relay (Fig. 7.6a, b). We see the cross section of the Buchholz relay, and a detail of the Buchholz relay.

In this scenario, the transformer disconnects before an incipient fault provokes a major fault, as the Buchholz relay provides an alarm. Happen a significant short-circuit in the transformer, the pressure within the primary tank increases, resulting in the rapid transfer of the cooling oil from the primary tank (illustrated as "main tank" in Fig. 7.6b) to the expansion tank (referred to as "conservator tank" in Fig. 7.4). Note that the oil pressure originating from the main tank, showed as "main tank" (Fig. 7.4) will trigger the circuit breaker and the alarm. Regarding its external appearance, it is important to note that, while the operating principle remains the same, the visual design of a power transformer can vary. It is common to find bigger power transformer at substations (Fig. 7.7).

Transformers exhibit diverse characteristics contingent upon their operational parameters, including power, weight, voltage, current and others. The following figures illustrate these aspects. Other arrangements, like a distribution pole transformer does not have an expansion tank (Fig. 7.8).

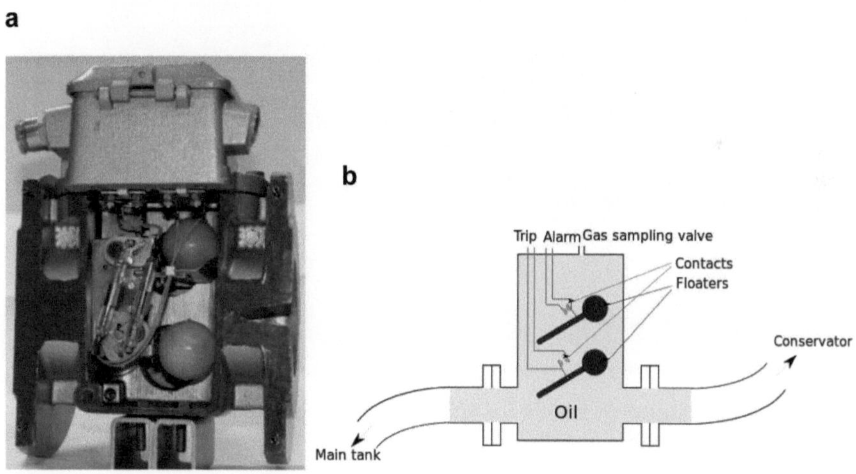

Fig. 7.6 Duplicate **a** Buchholz relay **b** Buchholz relay detail. Original from Fluppe37 and Psemdel at nl.wikipedia. *Source* From wikicommons under 3.0 CC use license [5, 6]

Fig. 7.7 Power Transformer 10 kVA. Original from Tirupati Transformers Pvt Ltd. *Source* From wikicommons under 4.0 CC use license [7]

Fig. 7.8 Pole type distribution transformer without an expansion tank. *Source* From wikicommons under free to use license [8]

7.4 Ideal Single-Phase Power Transformer

Before the analysis of how real transformers work, we first examine the ideal transformer model. We consider a transformer to be ideal when the following four conditions are met:

Firstly, the magnetic circuit must be unsaturated, meaning that the B-H curve displays no saturation. Secondly, there must be no iron losses within the magnetic circuit, and no hysteresis cycle. Thirdly, it is necessary that the transformer windings have zero resistance to ensure that there are no Joule effect losses. Finally, it is imperative that no stray flux exists, as this will ensure that all the flux passes through the magnetic circuit.

7.4.1 Ideal Single-Phase No-Load Power Transformer

A transformer is under no-load condition when, being fed by its primary, no load connects to the secondary.

In case Z_L remains null (Fig. 7.9) shows a single-phase in no-load condition transformer. If we apply sinusoidal voltage $V_1(t)$ through the primary it causes the circulating flux $\Phi_m(t)$ through the iron core.

In compliance with Hopkinson's law, this flux $\Phi_m(t)$ causes in turn a current $i_{10}(t)$ through the primary coils.

Thus, according to Hopkinson's law, it is true that:

$$N_1 \cdot i_{10}(t) = \Phi_m(t) \cdot \Re \tag{7.1}$$

It is of paramount importance to comprehend the waveforms that appear in the case of a real transformer. It is essential to keep in mind the order of imposition of the magnitudes in the primary of the transformer is:

$$v_1(t) \rightarrow \Phi_m(t) \rightarrow i_{10}(t) \tag{7.2}$$

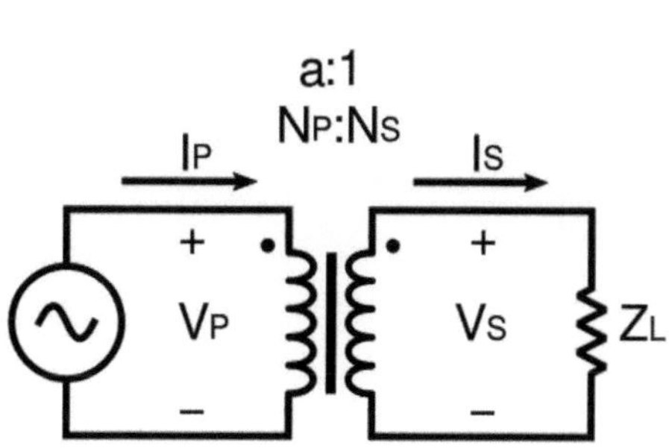

Fig. 7.9 Transformer under no-load condition. *Source* From wikicommons under free to use license [9]

According to these premises, there is no leakage flux in an ideal transformer operating in vacuum. The magnetic core of the transformer and the primary and secondary turns enclose the flux Φ_m (t) in the primary circuit.

7.4.2 Kirchhoff's Second Law

Recalling Kirchhoff's second law, the sum of all the voltage drops around the closed loop equals zero including the voltage drop across the impedances (Fig. 7.10).

$$U = U_1 + U_2$$

By applying Kirchhoff's second law to the primary electrical circuit results in Faraday's and Lenz's law. The negative sign in Faraday's Law (Lenz's law) indicates the polarity of the induced emf.

The current that the induced emf causes travels in the direction that creates a magnetic field that opposing the change in the original flux through the circuit. Recalling Faraday's law:

$$\varepsilon = -N_1 \frac{d\Phi_m(t)}{dt} \tag{7.3}$$

It is coupled with Lenz's law.

Assuming that the resistance of the loops in an ideal transformer is zero, it results that:

$$v_1(t) - N_1 \frac{d\Phi_m(t)}{dt} = i_{10}(t) \cdot 0 \tag{7.4}$$

For the sake of simplicity in nomenclature, we now employ the same notation (Fig. 7.10):

$$v_1(t) = N_1 \frac{d\Phi_m(t)}{dt} = e_1 \tag{7.5}$$

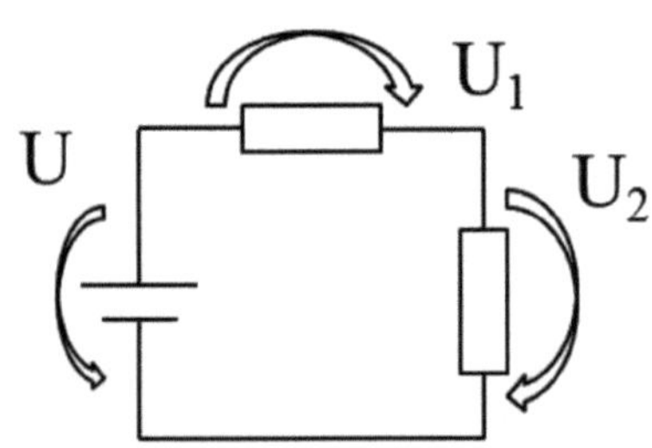

Fig. 7.10 Kirchhoff's second law. *Source* Own elaboration

Question If v_1 (t) is sinusoidal, what Φ m (t) be like?

From (Eq. 7.5), we conclude that if we apply sinusoidal voltage, the flux will also be sinusoidal. This flux $\Phi_m(t)$ is also responsible for inducing voltage e_2 in the turns of the secondary. For this reason, we usually name it mutual or useful flux.

Recalling (Eq. 7.6), then we can rewrite (Eq. 7.2) as:

$$\Phi_m(t) = \Phi_0 \cdot cos\omega t \tag{7.6}$$

The expressions of e1 and e2 become:

$$e_1 = N_1 \cdot \frac{d\Phi_m}{dt} = -N_1 \cdot \Phi_0 \cdot \omega \cdot sen\omega t = -N_1.\Phi_0.2\pi \cdot f \cdot \cos\left(\omega t - \frac{\pi}{2}\right) \tag{7.7}$$

$$e_2 = N_2 \cdot \frac{d\Phi_m}{dt} = -N_2 \cdot \Phi_0 \cdot \omega \cdot sen\omega t = -N_2 \cdot \Phi_0 \cdot 2\pi \cdot f \cdot \cos\left(\omega t - \frac{\pi}{2}\right) \tag{7.8}$$

Recalling that:

$$\omega = 2\pi \cdot f$$

$$senx = \cos\frac{\pi}{2} - x$$

$$\frac{d}{dx}(cosx) = -senx$$

If we use the effective values, we have:

$$E_1 = \frac{2\pi}{\sqrt{2}} \cdot N_1 \cdot \Phi_0 \cdot f = 4,44 \cdot N_1 \cdot \Phi_0 \cdot f \tag{7.9}$$

$$E_2 = \frac{2\pi}{\sqrt{2}} \cdot N_2 \cdot \Phi_0 \cdot f = 4,44 \cdot N_2 \cdot \Phi_0 \cdot f \tag{7.10}$$

The quotient of these rms values defines the transformer transformation ratio:

$$\text{Transformation ratio} = a = \frac{E_1}{E_2} = \frac{4,44 \cdot N_1 \cdot \Phi_0 \cdot f}{4,44 \cdot N_2 \cdot \Phi_0 \cdot f} = \frac{N_1}{N_2} \tag{7.11}$$

7.4.3 True No-Load Transformer

If we eliminate the four assumptions under which we develop the ideal transformer model (Fig. 7.9) we have the real transformer. We have the case of the real no-load transformer (Fig. 7.11). In it $\Phi\sigma_1$ represents the leakage flux that happens in the primary. In the absence of a load connected to the secondary winding, the transformer is in a state of no-load condition. Consequently, there is no dispersion or leakage flux. This is due to the fact that no current flows through the turns of the circuit. In this case i_2 equals zero because a load is not connected.

Considering the leakage flux, $\Phi_{\sigma 1}$, which is equivalent to eliminating the last assumption of the ideal transformer, we have that:

$$e_1 = N_1 \frac{d(\Phi_h + \Phi\sigma 1)}{dt} \tag{7.12}$$

$$e_2 = N_2 \frac{d\Phi_h}{dt} \tag{7.13}$$

In the secondary coil, there is no dispersion or leakage flux. This is because no current flows through its turns. Eliminating the third assumption of the ideal transformer (the transformer coils are of zero resistance) and applying Kirchhoff's second law to the primary circuit, we obtain:

$$v_1 - e_1 = R_1 \cdot i_{10}$$

where R_1 is the resistance of the primary coil. Substituting (Eq. 7.12) in to this expression and rearranging terms, we have that:

$$v_1 - N_1 \cdot \frac{d(\Phi_h + \Phi_{\sigma 1})}{dt} = R_1 \cdot i_{10} \tag{7.14}$$

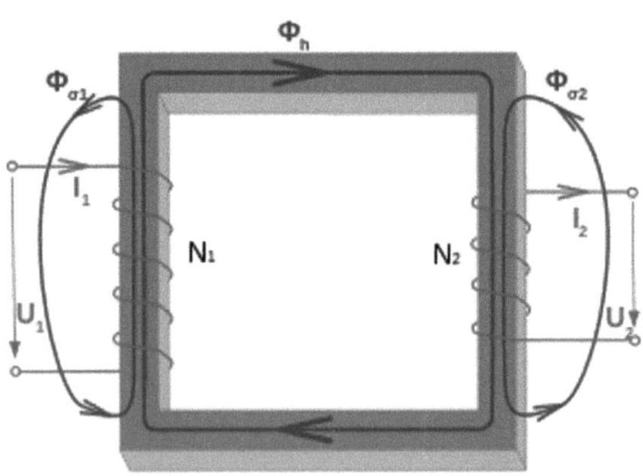

Fig. 7.11 Basic transformer. *Source* From wikicommons under free to use license [10]

$$v_1 - N_1 \cdot \frac{d\Phi_h}{dt} = R_1 \cdot i_{10} + N_1 \cdot \frac{d\Phi_{\sigma 1}}{dt} \tag{7.15}$$

Recalling that for an inductor:

$$v = L\frac{di}{dt}$$

where, V is the Voltage across the inductor, L is the inductance in Henries, $\frac{\mathrm{di}}{\mathrm{dt}} = 0$ is the instantaneous rate of current change (amps per second).

It is reasonable to assume a linear behavior because the dispersion or leakage flow occurs in the air. Therefore, we have that:

$$N_1 \cdot \frac{d\Phi_d}{dt} = L_1 \cdot \frac{di_{10}}{dt} \tag{7.16}$$

where $\mathrm{L_1}$ is the constant self-induction coefficient. (It is constant precisely because the leakage flow path is mainly air). Therefore, rearranging these equations, we have:

$$v_1 - N_1 \cdot \frac{d\Phi_m}{dt} = R_1 \cdot i_{10} + L_1 \cdot \frac{di_{10}}{dt} \tag{7.17}$$

Question What conclusions does this equation have?

Knowing that $\mathrm{v_1 - e_1}$ equals 0 in the ideal transformer,

What do we conclude from the point of view of the value of the characteristic parameters of the $\mathrm{N_1}$ coil?

Now, we know that (Eq. 7.18) shows that a real coil is like connect an ideal coil and the impedance ($\mathrm{Z_1}$) that connects in series.

$$\mathrm{v_1 - e_1 = R_1 \cdot i_{10} + L_1 \cdot \frac{di_{10}}{dt}} \tag{7.18}$$

We can model the effects of stray flux ($\mathrm{X_1}$), Joule losses ($\mathrm{R_1}$) as well others of real inductances (Fig. 7.12).

Recalling Eqs. 7.9, and 7.10 of real transformers:

$$E_1 = \frac{2\pi}{\sqrt{2}} \cdot N_1 \cdot \Phi_0 \cdot f = 4,44 \cdot N_1 \cdot \Phi_0 \cdot f \tag{7.19}$$

$$E_2 = \frac{2\pi}{\sqrt{2}} \cdot N_2 \cdot \Phi_0 \cdot f = 4,44 \cdot N_2 \cdot \Phi_0 \cdot f \tag{7.20}$$

$$\text{Transformation ratio} = a = \frac{E_1}{E_2} = \frac{4,44 \cdot N_1 \cdot \Phi_0 \cdot f}{4,44 \cdot N_2 \cdot \Phi_0 \cdot f} = \frac{N_1}{N_2} \tag{7.21}$$

If we apply a sinusoidal voltage to the transformer, the flux will be sinusoidal only if we keep the two assumptions of the ideal transformer: magnetic circuit not under

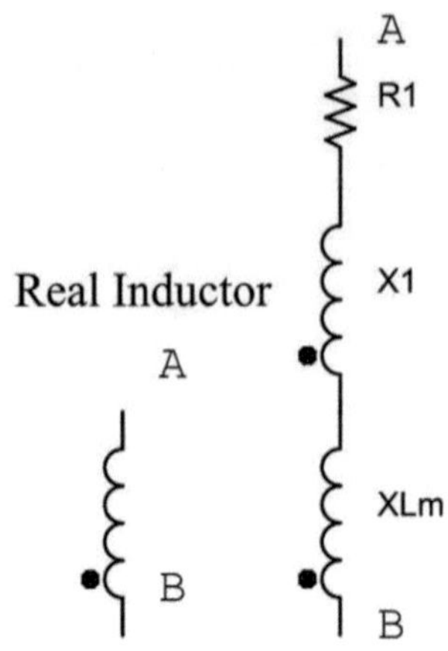

Fig. 7.12 Real inductances. *Source* Own elaboration

saturation and no iron losses in the magnetic circuit. We can represent the continuous hysteresis loop in the core of a 50 Hz toroidal core transformer in no load operation (Fig. 7.13).

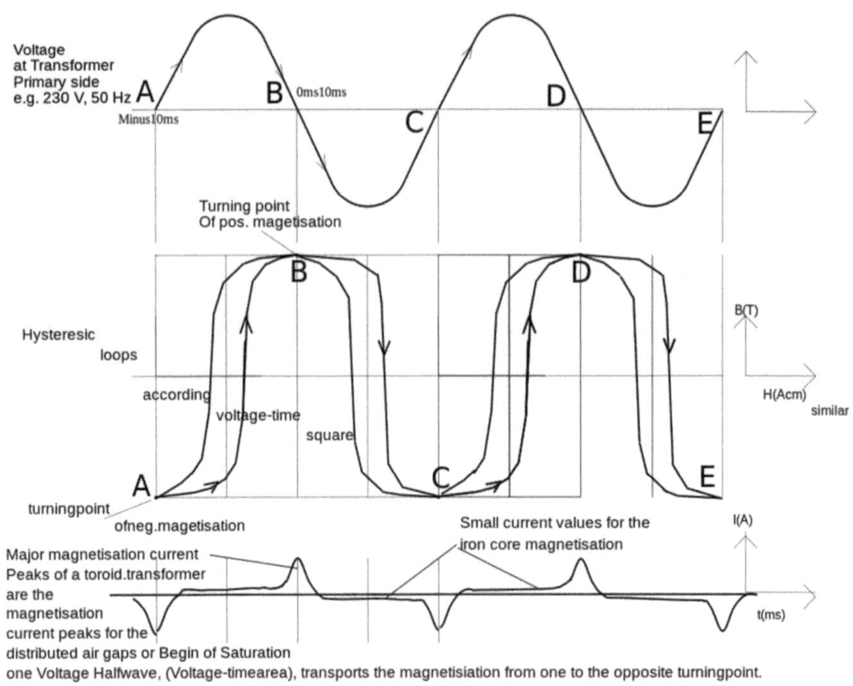

Fig. 7.13 Continuous Hysteresis loop in the core of a 50 Hz Toroidal core Transformer in permanent no-load situation. *Source* From wikicommons under free to use license [11]

However, the vacuum intensity will not be sinusoidal (note the "deformed" waveform). This is because of the influence of the hysteresis cycle (the cycle you see above on the right). A transformer is loaded when, being fed by its primary, a load connects to the secondary. Then, we have alternative configurations for transformers with different magnetic circuit (Fig. 7.14).

Under these conditions, the current flows through the primary and secondary coils (I_1, and I_2). The dispersion or leakage flux flows in both windings. The induced B field in a loop of wire will oppose the change in magnetic flux through the loop. The induced field will oppose any attempts to increase the flux through a loop. Any attempt to reduce the flux through a loop will be replaced by the induced field. Now, we represent the induced field B in a loop (Fig. 7.15).

According to Lenz's law to the secondary circuit, remember that the current in its windings must oppose to the flux variation that originates it. For this reason, the magnetomotive force that appears in the secondary (N_2–I_2) opposes the magnetomotive force that appears in the primary (N_1–I_1). This is because it is the latter (the

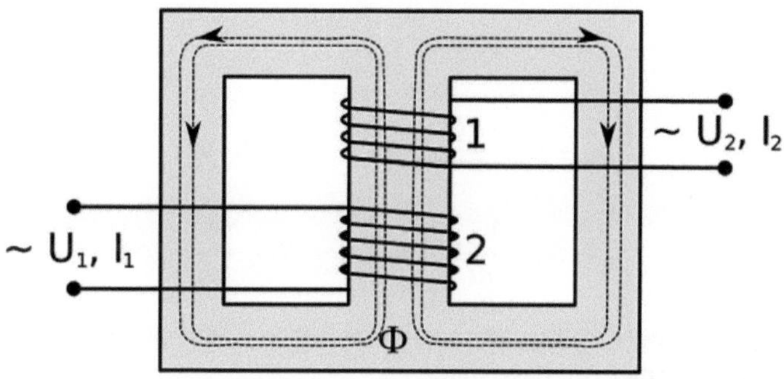

Fig. 7.14 Loaded transformer. *Source* From wikicommons under free to use license [12]

Fig. 7.15 Induced field B in a loop. *Source* Own elaboration

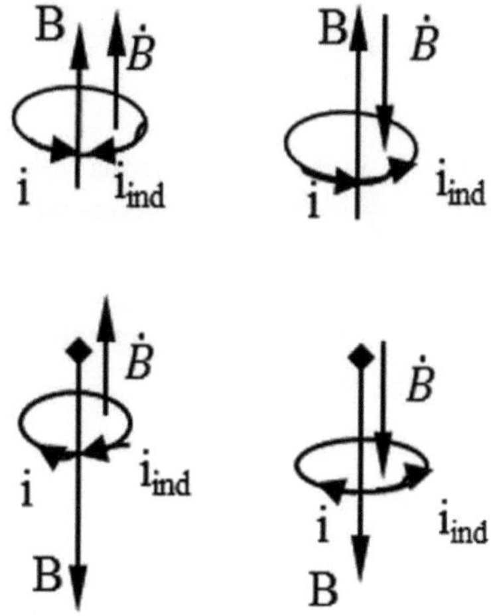

intensity) that determines the mutual flux direction. Remember that the current is a vector and, as such, has a direction (which does not occur with the number of turns). We must also keep in mind that dispersion fluxes in both coils are due to the currents that flow through them.

As a result, the leakage flux in the primary Φ_{d1} has the same direction as the mutual flux Φ_{dm} and the leakage flux in the secondary Φ_{d2} has a different direction than the mutual flux. Therefore, under load, the expressions of the electromotive forces in the primary and the secondary windings are:

$$e_1 = N_1 \cdot \frac{d(\Phi_{dm} + \Phi_{d1})}{dt} \tag{7.22}$$

$$e_2 = N_2 \cdot \frac{d(\Phi_{dm} - \Phi_{d2})}{dt} \tag{7.23}$$

Applying Kirchhoff's second law to the primary circuit and developing the expression:

$$v_1 - e_1 = R_1 \cdot i_1$$

$$v_1 - N_1 \cdot \frac{d(\Phi_{dm} + \Phi_{d1})}{dt} = R_1 \cdot i_1 \tag{7.24}$$

$$v_1 - N_1 \cdot \frac{\Phi_{dm}}{dt} = R_1 \cdot i_1 + N_1 \cdot \frac{d\Phi_{d1}}{dt} \tag{7.25}$$

If we call the induced voltage of the mutual flux as e_1 and proceeding like we did for the ideal transformer, we have that:

$$v_1 - e_1 = R_1 \cdot i_1 + L_1 \cdot \frac{di_1}{dt} \tag{7.26}$$

$$v = L\frac{di}{dt}$$

where V is the instantaneous voltage across the inductor, L is the inductance in Henrys, $\frac{di}{dt}$ is the instantaneous rate of current change (amps per second).

Recalling the instantaneous voltage induced in a coil:

$e = L - di/dt$ where L_1 is a constant self-induction coefficient. (It is constant precisely because the leakage flux path is primarily air).

It also follows that:

$$E_1 = 4,44 \cdot N_1 \cdot \Phi_0 \cdot f$$

and:

$$\mathrm{V}_1 \approx \mathrm{E}_1$$

Similarly, applying Kirchhoff's second law to the secondary circuit and substituting in (Eq. 7.24), we have:

$$e_2 - v_2 = R_2 \cdot i_2 \tag{7.27}$$

$$N_2 \cdot \frac{d(\Phi_m - \Phi_{d2})}{dt} - v_2 = R_2 \cdot i_2 \tag{7.28}$$

$$N_2 \cdot \frac{d\Phi_m}{dt} - v_2 = R_2 \cdot i_2 + N_2 \cdot \frac{d\Phi_{d2}}{dt} \tag{7.29}$$

Using the same nomenclature criteria as in the primary circuit:

$$e_2 - v_2 = R_2 \cdot i_2 + L_2 \cdot \frac{di_2}{dt}$$

As in the primary coil, the following is still true:

$$E_2 = 4,44 \cdot N_2 \cdot \Phi_0 \cdot f$$

Then:

$$\mathrm{V}_2 \approx \mathrm{E}_2$$

We use these expressions to compute the transformation ratio, one can determine the ratio of primary to secondary voltages. Then, we define the transformation ratio as:

$$\text{Transformer ratings} = \frac{V_1}{V_2} \approx \frac{E_1}{E_2} = \frac{N_1}{N_2} = a$$

Conversely, the following relationship holds true in the magnetic circuit (Hopkinson's law):

$$N_1 \cdot i_1 - N_2 \cdot i_2 = \Phi_m \cdot \Re \tag{7.30}$$

Given the extremely small reluctance, it is possible to state the following without significantly misstating anything:

$$N_1 \cdot i_1 \approx N_2 \cdot i_2$$

Thus, the ratio of primary and secondary currents is the inverse of the ratio of voltages:

$$\frac{I_1}{I_2} = \frac{N_2}{N_1} = \frac{1}{\frac{N_1}{N_2}} = \frac{1}{a} \tag{7.31}$$

7.5 Ratings

Question What are transformer ratings?

The ratings of a power transformer are the voltage, current and other design values. Electrical machines and transformers specifically cannot operate under all conditions. Every one of them is made to withstand specific boundary conditions.

Question Then, how do we design and engineer transformers and other electrical machinery? What are nominal conditions?

The settings of a power transformer that allow operating indefinitely are the rated conditions. These conditions imply nominal values of voltage, current and power.

Rated voltage

This is the voltage that the transformer can withstand for an indefinite period. Both the criterion of preventing the transformer from reaching saturation and the electrical insulators that employ in its construction determine its value. The nominal voltage value is also present in each transformer winding (Fig. 7.16). The current that the transformer windings can sustain for circulation is the rated current. Its value is determined by the area of the wires. Each transformer winding, primary or secondary has its own rated current value. We find the main values of a power transformer in the plate (Fig. 7.16).

Considering the existing relationships between voltages and currents and the fact that it would not make sense to build a transformer with windings prepared to withstand non-equivalent rated conditions in primary and secondary, we have that:

Fig. 7.16 Nameplate with the values of the ABB transformer. *Source* From wikicommons under free to use license [13]

$$\frac{V_{1N}}{V_{2N}} = \frac{I_{2N}}{I_{1N}} = a$$

Consequently, the single-phase transformer's rated power is a single value for both primary and secondary, which is indicative of the transformer's overall characteristics. We can express it as:

$$S_N = V_{1N} \cdot I_{1N} = V_{2N} \cdot I_{2N} \tag{7.32}$$

Thus, the following is defined:

1. S_N Transformer apparent power rating
2. V_{1N} Rated primary voltage
3. V_{2N} Nominal secondary voltage
4. I_{1N} Rated primary current
5. I_{2N} Rated secondary current.

The transformer nameplate itself lists these values. A transformer usually works at its rated voltage, although sometimes loads can be variable.

7.5.1 *Voltage Regulation by Varying Transformer Windings*

The number of turns on the primary and secondary loops of a transformer determines the turns ratio among other things. If this number is invariable, we can only use one turns ratio Therefore, for an input voltage (V_1) to the transformer, the output voltage (V_2) is univocally defined for a given load (Z_C) (Fig. 7.17).

As a result, the transformer satisfies the expressions below:

$$\underline{V}_2 = \underline{E}_2 - \underline{I}_2 \cdot (R_2 + jX_2) \tag{7.33}$$

$$\underline{V}_1 = \underline{E}_1 - \underline{I}_1 \cdot (R_1 + jX_1) \tag{7.34}$$

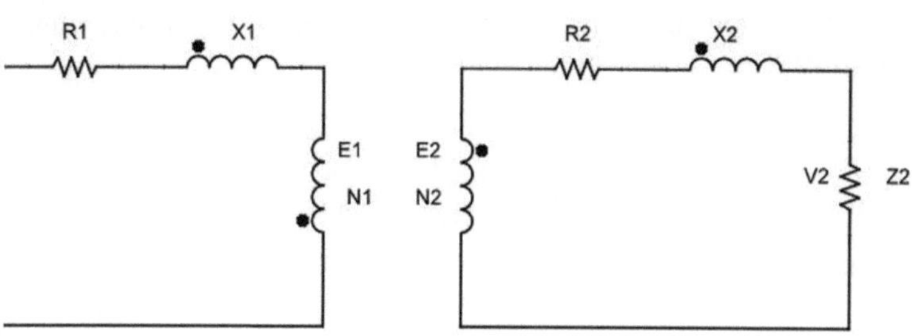

Fig. 7.17 Equivalent circuit of a loaded transformer. *Source* Own elaboration

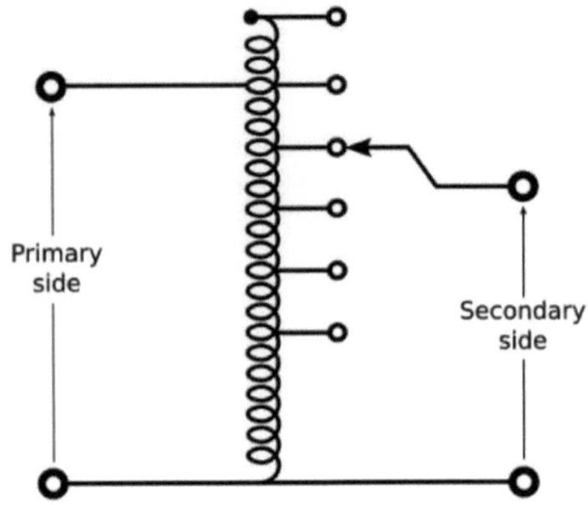

Fig. 7.18 Regulation on the secondary windings. *Original Source* From wikicommons under free CC 3.0 use license [14]

$$\frac{\underline{E}_1}{\underline{E}_2} = \frac{N_1}{N_2}$$

Operating with these three equations we obtain:

$$\underline{V}_2 = \frac{N_1}{N_2} \cdot \left[\underline{V}_1 - \underline{I}_1 \cdot (R_1 + jX_1)\right] - \underline{I}_2 \cdot (R_2 + jX_2) \tag{7.35}$$

This expression shows how the turns ratio affects the voltage V_2. In order to regulate this voltage, different taps can be established. We call them transformer taps. If so in a power transformer, turns ratio can change.

Question Where do we locate the transformer taps?

Depending on the manufacturer we can find the taps of a transformer on the primary, on the secondary, or on both windings at the same time. The tap changer can adopt different schemes (on the right) (Fig. 7.18). Only by moving the taps of the power transformer changes the turns ratio.

However, installing transformer taps on both windings at the same time is more complex, more expensive and does not provide any additional performance. It is interesting to identify a "real" tap changer where we see how they connect the taps (Fig. 7.19).

Although, there are alternative designs for tap changer transformer (Fig. 7.20).

Question Why do we put taps on transformers?

In answer to this question, we install taps to:

- Keep a constant secondary voltage against changes in the primary voltage.
- Keep the secondary voltage constant against load variations.

Question How many taps can a transformer have?

It is possible to set up any number of taps. However, from an economic point of view, it is advisable not to exceed the regulatory range. This is because the price of the transformer increases.

An example of the information we can find in a transformer plate is:

Fig. 7.19 Taps in a power transformer. Fluppe 37 at Dutch wikipedia. *Source* From wikicommons under CC 3.0 use license [15]

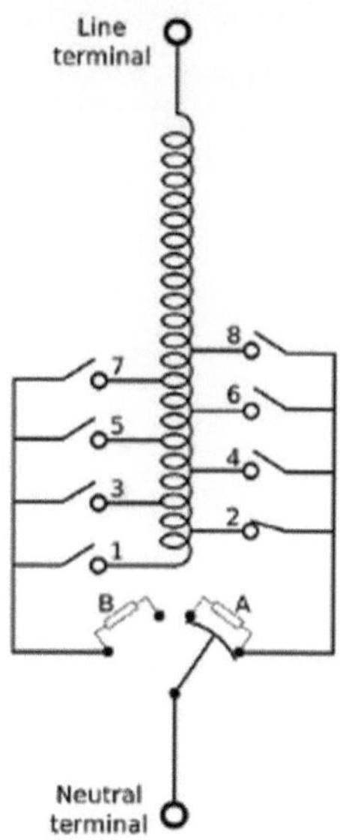

Fig. 7.20 An 8-tap changer transformer in a power transformer. *Original Source* From wikicommons under CC 3.0 use license [16]

$$30,000\text{V} \pm 2.5\% \pm 5\% \pm 5\% \; - 7.5\%/400 \text{ V}$$

This information belongs to a transformer with a nominal ratio of 30.000/400, with a primary regulation range from + 5% to −7.5% with 2.5% steps.

It is worth saying that devices regulate voltage in a no-loaded transformer, although under load conditions, it needs more sophisticated devices. These devices must guarantee the continuity of supply during the turn ratio change maneuver.

7.6 Transformer Efficiency

The transformer efficiency measures the amount of energy it delivers versus the energy needed to deliver it. We know that in both, the primary and the secondary, there are losses because of resistance of coils (Copper losses). It also appears leakage flux (Iron losses) (Fig. 7.21). We define the efficiency of a transformer as:

$$\eta = \frac{P_{u'til}}{P_{absorbed}} = \frac{P_{useful}}{P_{useful} + P_{losses}} = \frac{P_2}{P_2 + P_{Fe} + P_{Cu}} \quad (7.36)$$

Alt text: single line diagram illustrating the model of an ideal transformer w.

Suppose we have a transformer where we supply the rated voltage V_{1N} to the primary, and an inductive load that connects to the secondary, which absorbs a certain current, I_2.

In this case, the useful output power is:

$$P_2 = V_2 \cdot I_2 \cdot \cos\varphi.$$

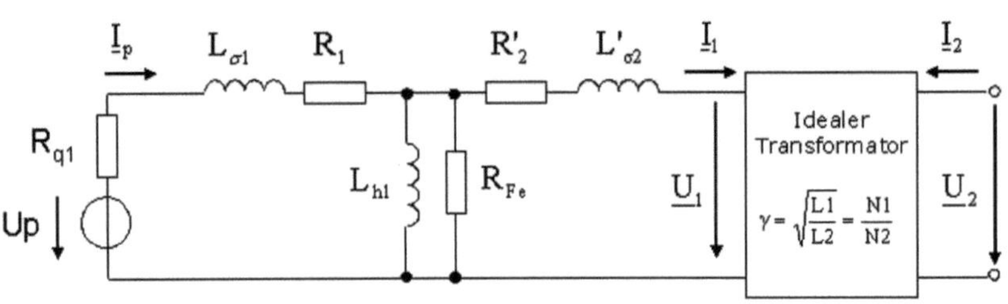

Fig. 7.21 Power transformer losses Pfe, and Pcu. *Source* From wikicommons under free to use license [17]

7.7 Operation of the Transformer at a Frequency Different from the Rated Frequency

This section analyzes what happens in a transformer if the frequency of the applied voltage decreases while keeping the rms value of the applied voltage constant.

$$V_1 = 4,44 \cdot N_1 \cdot \Phi_0 \cdot f$$

Under these conditions, the frequency decrease causes a proportional flow increase since the product of both must be constant. As previously seen, the expression of iron losses is:

$$P_{Fe} = P_F + P_H = K_F \cdot f^2 \cdot B^2_{ma'x} + K_H \cdot f \cdot B^x_{ma'x}(\text{Hysteresis losses}) \qquad (7.37)$$

$$\mathrm{P_{Fe}} : \text{Iron losses}$$

Because of hysteresis losses, the B^2max factor increases proportionally to the frequency decrease. Then, the iron losses increase. Note that for hysteresis losses, the induction increases to a factor x > 1. Remember also that the flux (Φ) equals the induction (B) times the section (S).

As a consequence of the above, the transformer may begin to saturate as the flux increases. On the other hand, if the frequency is increased while the rms voltage remains constant, the opposite effect occurs. In this case, both flux and iron losses decrease. It could appear that the second scenario is preferable. But there are disadvantages as well. This is because frequency increase provokes an increase in the value of the internal reactance as well (remember that $X = 2\pi fL$). As a consequence, the voltage drop that will occur in the transformer will be higher.

7.8 Operation of the Transformer at a Voltage Different from the Rated Voltage

$$V_1(\uparrow) = 4,44 \cdot N_1 \cdot \Phi_0 \cdot f$$

Now we analyze the case in which the rms value of the voltage supply increases, keeping the frequency constant. Under these conditions, we see a flux increase which leads to an increase in iron losses. These conditions mean that:

$$\mathrm{V_1 > V_{1N}}$$

Then, because:

$$f = cte \rightarrow \Phi_0 \uparrow \rightarrow P_{Fe} \uparrow$$

$$P_{Fe} = P_F + P_H = K_F \cdot f^2 \cdot B_{ma'x}^2 + K_H \cdot f \cdot B_{ma'x}^x \tag{7.38}$$

Then, we have to keep in mind that if we expose the insulators to a voltage higher than the design voltage, or if we apply a voltage higher than the rated voltage (at the same frequency), the transformer become at the risk of saturation, with the associated risks we discussed the previous section. It is worth saying that the transformer is not at risk if we apply a voltage lower than the rated voltage (at the same frequency).

Question What happen if we apply to the transformer a voltage lower than the nominal voltage (of the same frequency)?

In this case the flux is lower, its iron losses are lower, and there is no saturation risk. However, it cannot be used at the maximum power for which it was built.

7.9 Types of Three-Phase Transformers

Similarly in a three-phase transformer we also mainly find: Core, Low Voltage windings, main channel, High Voltage winding. Depending on the design features, the three-phase transformer can have three independent magnetic cores (Fig. 7.22).

Other types of three-phase transformers use a single magnetic core to reduce magnetic core, reducing weight among others (Fig. 7.23).

We can arrange a three-phase transformer, using the same components: the core, primary, and secondary windings, and insulation, but with different design (Fig. 7.24).

By making the star or delta connections to the primary and secondary electrical windings, we obtain a three-phase transformer. To build the bench, three identical single-phase transformers must be used.

The main advantage of a three-phase core transformer is economical. It is important to highlight that a single housing must enclose all magnetic cores. Therefore, if

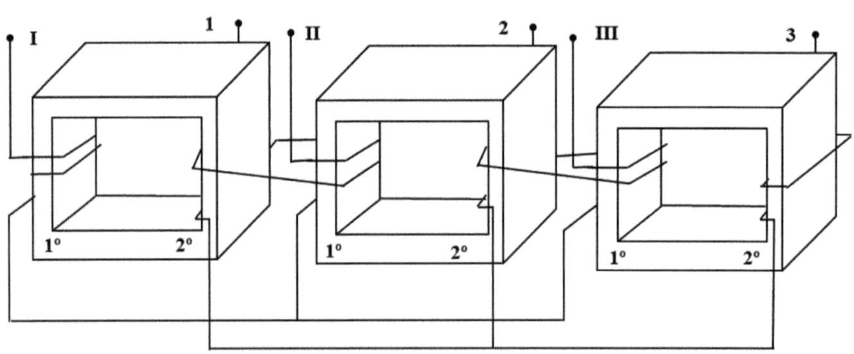

Fig. 7.22 Transformer power bank. *Source* Own elaboration

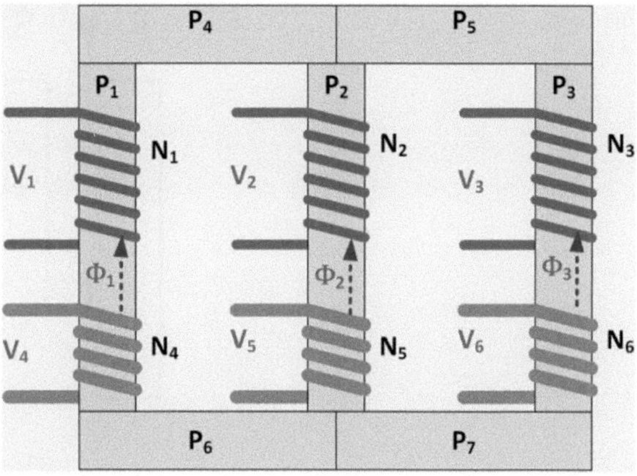

Fig. 7.23 Three-phase core transformer. *Source* From wikicommons under free to use license [18]

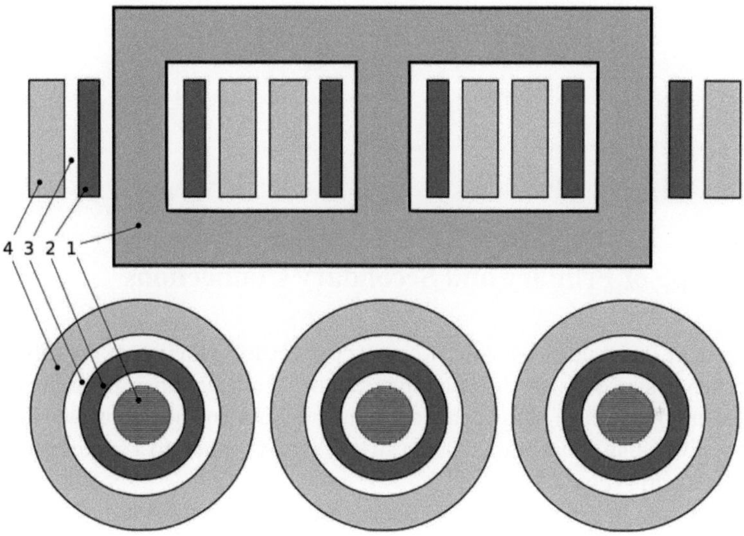

Fig. 7.24 Three-phase and single core transformer. *Source* From wikicommons under free to use license [19]

we use three different housings for three single-phase transformers, we employ more iron if we only use one that contains a three-phase transformer (even if it has a larger size). The disadvantage lies in having a spare unit for replacement in case of failure. This is because it is much cheaper to have a spare fourth single-phase transformer than to duplicate a three-phase transformer.

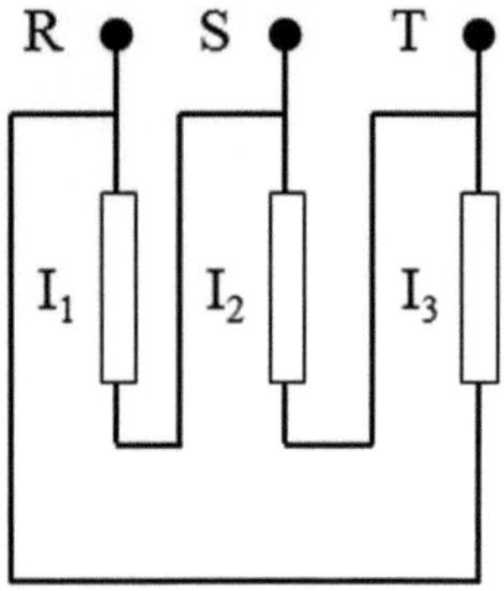

Fig. 7.25 Delta connection schematic. *Source* Own elaboration

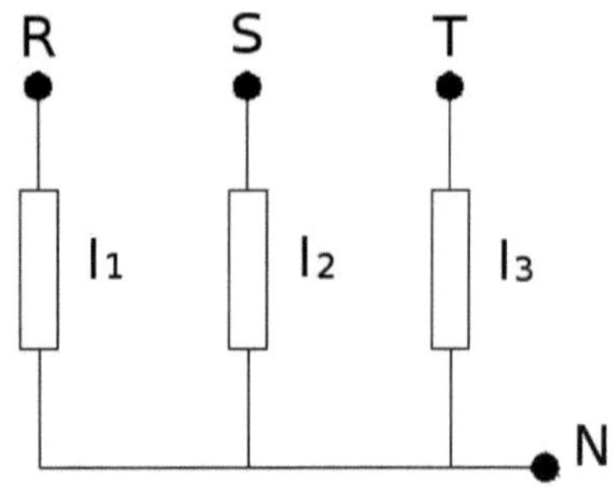

Fig. 7.26 Star connection schematic. *Source* From wikicommons under free to use license [20]

7.10 Type of Primary and Secondary Connections

As shown in the schemes below, we can connect the windings of a three-phase transformer in star, delta, or zigzag configuration, in both windings. In a delta winding connection scheme each phase connects to the previous and next as shown Fig. 7.25.

In a star winding connection scheme all the phases connect to a common terminal, known as neutral (Fig. 7.26).

Furthermore, in a zigzag winding connection scheme consisting of a mix of the previous schemes (Fig. 7.27).

7.11 Single-Phase Equivalent Circuit

Let's take the transformer shown in Fig. 7.28, which has its primary (high voltage) connected in delta and its secondary (low voltage) in star with the accessible neutral (Dyn).

As previously mentioned, in a star connection, the star always tries to equalize the current, and we get two voltage levels that can connect a single-phase load. It is expressed as:

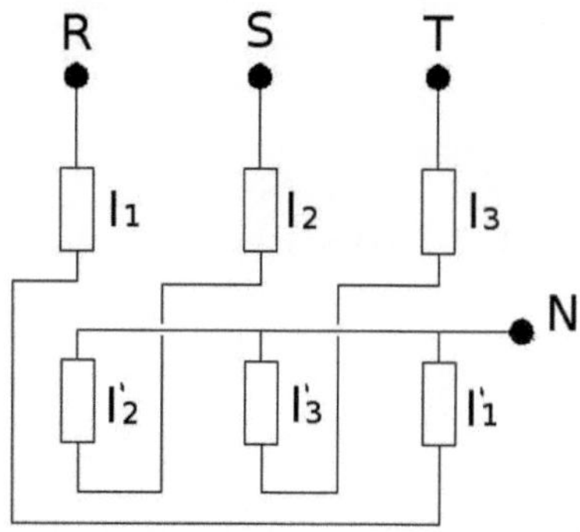

Fig. 7.27 Zigzag connection schematic. *Source* From wikicommons under free to use license [21]

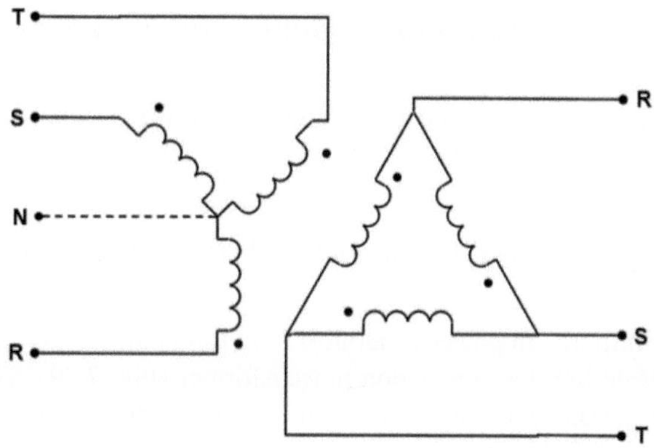

Fig. 7.28 Delta to star transformer connection. *Source* From wikicommons under free to use license [22]

$$IL = IP, \text{ and } V_L = \sqrt{3} \cdot V_P \tag{7.39}$$

And delta tries to equalize the voltage. It is used in equal voltage situations like motors etc. It is expressed as:

$$I_L = \sqrt{3} \cdot I_P \text{ and :}$$

$$V_L = V_P \tag{7.40}$$

Note that in the star connection, the current that flows through the line (I_L) is the same as the phase current (Ip). The current in the line (I_L) equals the phase current (I_P); while the line voltage (V_L) is the square root of 3 phase voltages (V_P).

In the delta connection, the line voltage (V_L) is the same as the phase voltage (V_P). The line current (I_L) is the square root of 3—phase voltage (I_P).

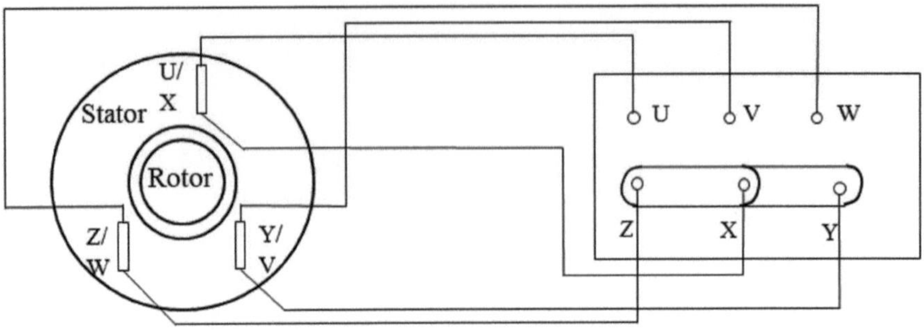

Fig. 7.29 Star generator connection. *Source* Own elaboration

7.12 Connection of Generators and Receivers

Next, we review the different ways that we use to connect generators and receivers.

7.12.1 *Star-Connection of the Transformer*

Similarly, in power generators, windings can be star-connected which slightly differ from the star connection in transformer (Fig. 7.29). The three free terminals of the generator windings connect to a common point, forming the neutral conductor (0). For a full comprehension of this connection, the generator connects to a receiver comprising three ohmic-inductive loads (Z_1, Z_2 and Z_3) that also connect together in a star configuration (Fig. 7.29).

7.12.1.1 Single or Phase Voltages

Each winding of the three-phase generator behaves as a single-phase generator, generating a voltage between its terminals called single or phase voltage (U_S): U_{10}, U_{20} and U_{30}.

7.12.1.2 Line Currents

The simple voltages remain applied to each of the receiver loads, appearing a current for each line conductor (I_L): I_1, I_2 and I_3. The sum of the three currents is the neutral return current I_N. At first, it might seem that the neutral conductor carries a large amount of electric current, but as we see from analyzing vector diagrams, in a balanced system where the charges are all equal (balanced charges) this current is zero. This could lead to erasing this wire under certain conditions.

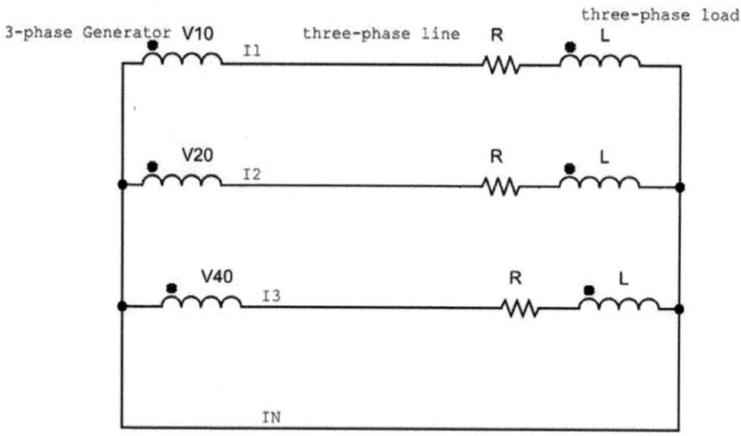

Fig. 7.30 Star-compound voltage $\vec{U}_{12}$. *Source* Own elaboration

7.12.1.3 Compound Voltages

They are the voltages that appear between two phases.

$$(U_C) : U_{12}, U_{23} \text{ and } U_{31}$$

We call these voltages as compound voltages and they result from single voltage addition. Only by applying Kirchhoff's laws and the vector diagram, we show that these voltages have the same rms value and that they are 120° out of phase regarding each other, as well as that (Fig. 7.30)

$$U_C = \sqrt{3} \cdot U_S$$

The following results after applying Kirchhoff's second law to each loop that appears between the simple and compound voltages (Fig. 7.31).

$$\vec{U}_{10} - \vec{U}_{12} - \vec{U}_{20} = 0 \quad (7.41)$$

$$\vec{U}_{12} = \vec{U}_{10} - \vec{U}_{20} \quad (7.42)$$

We can also write:

$$\vec{U}_{20} - \vec{U}_{23} - \vec{U}_{30} = 0 \quad (7.43)$$

$$\vec{U}_{23} = \vec{U}_{20} - \vec{U}_{30} \quad (7.44)$$

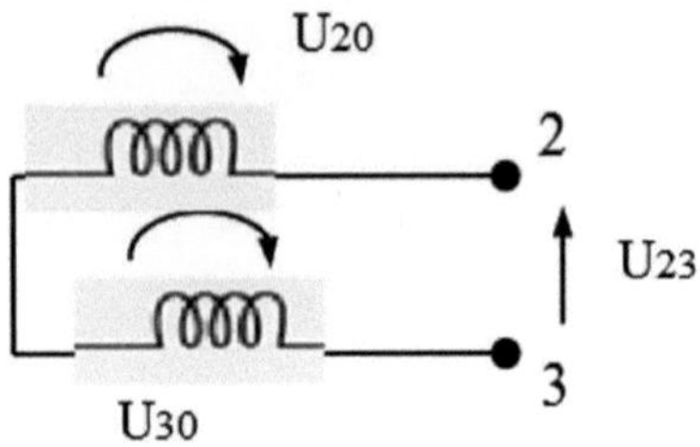

Fig. 7.31 Star-compound voltage $\vec{U}_{23}$. *Source* Own elaboration

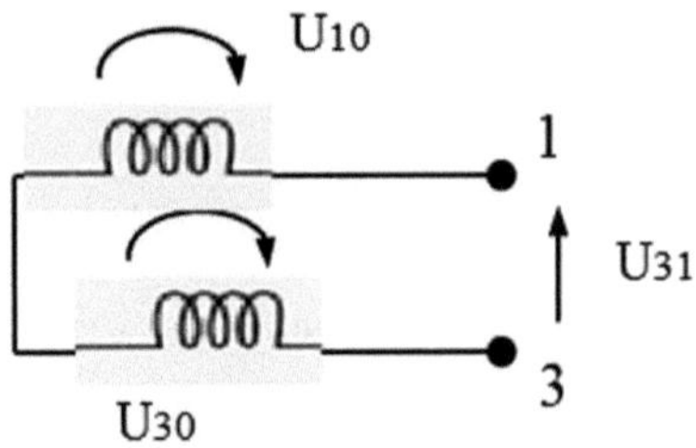

Fig. 7.32 Star compound voltage $\vec{U}_{31}$. *Source* Own elaboration

In the final loop we also have that:

$$\vec{U}_{10} - \vec{U}_{31} - \vec{U}_{30} = 0 \tag{7.45}$$

$$\vec{U}_{31} = \vec{U}_{30} - \vec{U}_{10} \tag{7.46}$$

Now we draw the vector diagram of the previous circuit (Fig. 7.32) following the simple voltages offset from each other by 120°. Now, we see the angle between the simple and compound voltages is 30º (Fig. 7.33).

Because an isosceles triangle appears we have that:

$$\alpha = \beta$$

$\vec{U}_{10} - \vec{U}_{12} - \vec{U}_{20} = 0$ Rearranging,

$$\vec{U}_{12} = \vec{U}_{10} - \vec{U}_{20}$$

Then:
$\vec{U}_{20} - \vec{U}_{23} - \vec{U}_{30} = 0$ Rearranging,

$$\vec{U}_{23} = \vec{U}_{20} - \vec{U}_{30}$$

And:
$\vec{U}_{10} - \vec{U}_{31} - \vec{U}_{30} = 0$ Rearranging,

Fig. 7.33 Vector diagram.
Source Own elaboration

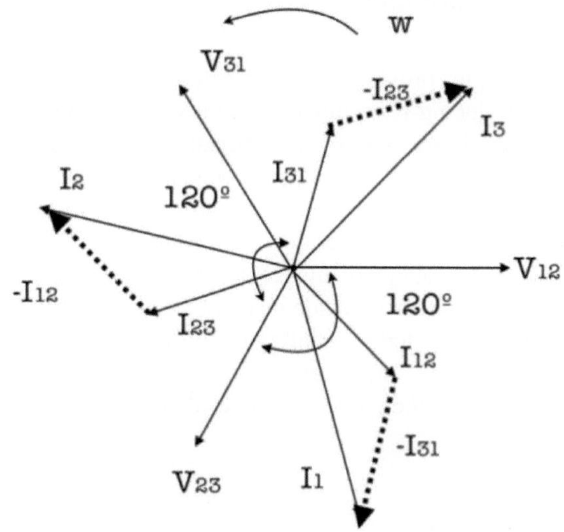

$$\vec{U}_{31} = \vec{U}_{30} - \vec{U}_{10}$$

In the previous triangle (Fig. 7.33), the following condition is true:

$$180^\circ = \alpha + \beta + 120^\circ \tag{7.47}$$

From which it follows that the angle is: $\alpha = 30^\circ$.

Using this relationship and using simple trigonometry we find the relationship that exists between simple and compound voltages. We express it in the (Eq. 7.41):

$$\cos 30^\circ = \frac{U_{12}/2}{U_{10}} \rightarrow U_{12} = 2 \cdot U_{10} \cdot \cos 30^\circ \tag{7.48}$$

$$U_{12} = 2 \cdot U_{10} \cdot \frac{\sqrt{3}}{2} = U_{10} \cdot \sqrt{3} \tag{7.49}$$

In conclusion, we have that:

$$U_C = \sqrt{3} \cdot U_S \tag{7.50}$$

Once we have a three-phase voltage system, we connect loads in delta, star, or even single-phase loads between phase and neutral or between phases.

7.13 Summary

This chapter provides a comprehensive overview of power transformers. It begins by introducing their fundamental role and identifying their key components. The text then delves into the operational principles, starting with an analysis of losses, protection systems, and the theoretical model of an ideal single-phase transformer, including its no-load operation and the application of Kirchhoff's laws.

The chapter covers critical performance aspects such as transformer ratings, efficiency, and voltage regulation methods. It also explains the consequences of operating a transformer outside its rated frequency or voltage. Finally, the focus shifts to three-phase systems, detailing the different types of three-phase transformers, their various primary and secondary connection schemes (such as Star connections), and how to model them using a single-phase equivalent circuit for simplified analysis.

References

1. Pedalito, Deutsch: Transformator der zur Stromversorgung eines Radios verwendet wurde, mit abgenommener Isolation und auseindergezogenem Eisenkern (2023). Accessed 24 Dec 2023. [Online]. https://commons.wikimedia.org/wiki/File:Trafo_auseinanderbauen_seitlich_auseinandergezogen.jpg
2. Wikipedia. Mains transformer produced by A/S Per Kure Norwegian Motor and Dynamofabrik (2011). Accessed 8 Dec 2023. [Online]. https://commons.wikimedia.org/wiki/File:Mains_transformer_produced_by_AS_Per_Kure.jpg
3. O. uploader was F. at nl.wikipedia, Nederlands (2007). Accessed 8 Dec 2023. [Online]. https://commons.wikimedia.org/wiki/File:Vermogentransformator_1.GIF
4. Wikipedia. Laminated core eddy currents 2.svg—ikipedia, la enciclopedia libre». Accessed 8 Dec 2023. [Online]. https://commons.wikimedia.org/wiki/File:Laminated_core_eddy_currents_2.svg
5. F. at D. Wikipedia, Een doorgezaagde Buchholz relais laat duidelijk de beide vlotters en de Reed-contacten zien (2007). Accessed 8 Dec 2023. [Online]. https://commons.wikimedia.org/wiki/File:Buchholz_2.JPG
6. Psemdel: Description du fonctionnement d'un relais Buchholz en cas d'alarme (2014). Accessed 8 Dec 2023. [Online]. https://commons.wikimedia.org/wiki/File:Buchholz_relay_contacts_alarm_fr.svg
7. T.T.P. Ltd, English: A distribution transformer of 700 KVA. 2022. Accessed 8 Dec 2023. [Online]. https://commons.wikimedia.org/wiki/File:Distribution_Transformer.png
8. Sarang, English: Transformer on pole, in Baracoa. 2017. Accessed 8 Dec 2023. [Online]. https://commons.wikimedia.org/wiki/File:BCA_Distribution_transformer_2.jpg
9. O.C.V. Wikimpan, English: A depiction of an ideal transformer with a source and a load (2016). Accessed 8 Dec 2023. [Online]. https://commons.wikimedia.org/wiki/File:Ideal_transformer.svg
10. MitjaStachowiak, Deutsch: Ein realer Transformator mit Wicklungen und magnetischen Flüssen (2015). Accessed 24 Dec 2023. [Online]. https://commons.wikimedia.org/wiki/File:Modell_eines_realen_Transformators.svg
11. Emeko, Deutsch: Der Verlauf der Netzspannung beeinflusst die Magnetisierung, also die Flussdichte B und erzeugt einen Strom dessen Verlauf und Höhe von der Hysteresekurve abhängt (2014). Accessed 8 Dec 2023. [Online]. https://commons.wikimedia.org/wiki/File:Voltage_hysteresis_current_en.svg

12. Boom1x, Русский: Схема трансформатора. 2007. Accessed 24 Dec 2023. [Online]. https://commons.wikimedia.org/wiki/File:Transformator_scheme_ru.svg
13. Ordercrazy, Deutsch: Kurzwellensendeanlage Wertachtal - Rückbau im April 2015, Typenschild an einem Transformator der ABB-Trafo-BB. 2015. Accessed 8 Dec 2023. [Online]. https://commons.wikimedia.org/wiki/File:Kurzwellensendeanlage_Wertachtal_(004).jpg
14. BillC, English: An autotransformer showing taps on the secondary side. These provide a series of voltages approximately equal to 40%, 55%, 70%, 85%, 100% and 115% of the primary side voltage. 2006. Accessed 8 Dec 2023. [Online]. https://commons.wikimedia.org/wiki/File:Tapped_autotransformer.svg
15. F. at D. Wikipedia, Het aktief deel van een nl:autotransformator met 3 enkelfasige lastschakelaars. Deutsch: Drei einphasige Stufenschalter am Transformatordeckel montiert; die Regelableitung ist bereits angeschlossen (2007). Accessed 8 Dec 2023. [Online]. https://commons.wikimedia.org/wiki/File:3_OLTC.jpg
16. Wikipedia. Tap changer. Wikipedia. 30 September 2020. Accessed 8 Dec 2023. [Online]. https://en.wikipedia.org/w/index.php?title=File:Tap_changer.svg&oldid=981067150
17. M. Lenz, English: transformer equivalent circuit. Accessed 26 Dec 2023. [Online]. https://commons.wikimedia.org/wiki/File:Trafo-ESB.png
18. Wikipedia. Constant English: Three phase transformer with windings and permeance elements for use as an example with the gyrator-capacitor approach article (2019). Accessed 7 Dec 2023. [Online]. https://commons.wikimedia.org/wiki/File:Gyrator-Capacitor_Model_Example_Three_Phase_Transformer.png
19. Dermartinrockt, English: Principel of a three phase transformer with 3/0 core (2019). Accessed 7 Dec 2023. [Online]. https://commons.wikimedia.org/wiki/File:Transformer_Kern_und_Wicklungen.svg?uselang=de
20. Me, English: Star connection (Y) schematic in three-phase power networks (2007). Accessed 7 Dec 2023. [Online]. https://commons.wikimedia.org/wiki/File:Three-phase_star_schematic.svg
21. N. machine-readable author provided G. assumed, Zig-Zag connection schematic diagram (2006). Accessed 7 Dec 2023. [Online]. https://commons.wikimedia.org/wiki/File:ZigZag_connection_schematic.jpg
22. Wikipedia. Transformer types. Accessed 7 Dec 2023. [Online]. https://es.wikipedia.org/w/index.php?title=Tipos_de_transformador&oldid=154901321

Chapter 8
Connection of the Receivers

Abstract In electric systems is very important to understand how to connect the receivers in order to establish certain values of voltages, currents or other variables in the wires of the electrical system that they are connected to. After analyzing this point, we review the power that we find in the three-phase system, and study its relationship with the motor efficiency.

Keywords Star and delta connection · Power in three-phase system · Motor efficiency

8.1 Introduction

Question How do we connect loads to a three-phase system?

We connect three-phase loads to a three-phase system in either a delta or star configuration. We connect single phase loads between a phase and neutral or between different phases. In most cases, these loads are three-phase balanced loads. This includes three-phase motors and furnaces.

Single-phase loads include lamps and various types of single-phase receivers. In this situation, it is better to split the single-phase loads between the phases to avoid an unbalanced system. In each phase, an imbalance might result in destabilizing currents that would affect the system's stability.

Next, we look at the three-phase loads that connects in either both star and delta configurations. We also verify that, for balanced loads the power output for these connections is always similar. We express the power of a three-phase system as:

$$P = \sqrt{3} \cdot U_C \cdot I_L \cdot \cos\varphi \tag{8.1}$$

where P is the active power of the three-phase load, U_C is the composite voltage, I_L is the line current and cosφ is the power factor of the load.

D. Borge-Diez et al., *Fundamentals of Electrical Power Engineering*, Lecture Notes in Electrical Engineering 1540, https://doi.org/10.1007/978-3-032-18904-2_8

8.2 Star-Balanced Load

Now on, we analyze a three-phase voltage system with three star-connected loads (Fig. 8.1). Taking into account that it is a balanced nature system.

Then, we write for each of the loads:

$$Z_1\angle\varphi_1 = Z_2\angle\varphi_2 = Z_3\angle\varphi_3$$

Applying Ohm's law to each of these loads, we find the currents flowing through each phase:

$$\vec{I_1} = \frac{\vec{U_{10}}}{Z_1\angle\varphi_1} \quad (8.2)$$

$$\vec{I_2} = \frac{\vec{U_{20}}}{Z_2\angle\varphi_2} \quad (8.3)$$

$$\vec{I_3} = \frac{\vec{U_{30}}}{Z_3\angle\varphi_3} \quad (8.4)$$

Since the voltage is 120° out of phase, the system of currents is also 120° out of phase with each other, and has an angle φ regarding to each of their corresponding voltage. Since the impedance in the line and the applied voltage are equal, so does the current.

$$|I_1| = |I_2| = |I_3| = I_L \quad (8.5)$$

As we see in the vector diagram Fig. 8.4, it satisfies the following condition:

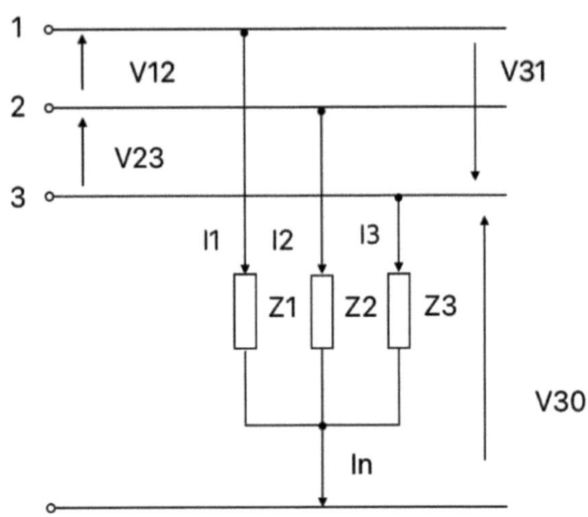

Fig. 8.1 Star balanced load. *Source* own elaboration

$$\vec{I_1} + \vec{I_2} + \vec{I_3} = \vec{I_N} = 0 \tag{8.6}$$

Question Under these circumstances, is it possible to eliminate the neutral conductor?

If this is the case, it is possible to create an artificial neutral at the most common point of the star-connected loads, which makes it possible to maintain a single voltage between the phases and the neutral without using the neutral conductor. It is important to keep in mind that this is only possible in balanced load scenarios.

8.2.1 Power of the Three-Phase System

To determine the power that a load dissipates, simply sum the power that the three single-phase loads dissipate individually, namely:

$$P = U_{10} \cdot I_1 \cdot cos\varphi_1 + U_{20} \cdot I_2 \cdot cos\varphi_2 + U_{30} \cdot I_3 \cdot cos\varphi_3 \tag{8.7}$$

Question How much power will the load consume?

The power of a three-phase system in balanced systems is three times that of a single phase. Phase and line currents for star loads are the same, as are single voltages, power factors, and phase currents. Then, we have:

$$P = 3 \cdot U_S \cdot I_L \cdot cos\varphi \tag{8.8}$$

And as we know:

$$U_S = \frac{U_C}{\sqrt{3}} \tag{8.9}$$

Then, substituting Eq. 8.9 in Eq. 8.8:

$$P = 3 \cdot \frac{U_C}{\sqrt{3}} \cdot I_L \cdot \cos\varphi \tag{8.10}$$

And finally, we have that:

$$P = \sqrt{3} \cdot U_C \cdot I_L \cdot \cos\varphi \tag{8.11}$$

P is the active power of the three-phase load, U_C is the compound voltage, I_L is the line current, and cos φ is the load power factor. Then, it appears the power triangle (Fig. 8.2).

Using simple trigonometry, we have:

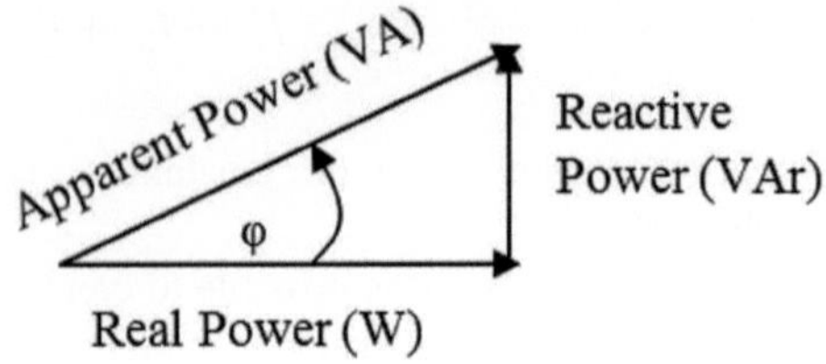

Fig. 8.2 Power triangle. *Source* own elaboration

$$S = \sqrt{P^2 + Q^2} \quad (8.12)$$

$$S = V \cdot I \quad (8.13)$$

$$P = V \cdot I \cdot \cos\varphi \quad (8.14)$$

$$Q = V \cdot I \cdot sen\varphi \quad (8.15)$$

In order to find the reactive, and the apparent power, we proceed similarly to the active power. Then, we have that:

$$Q = \sqrt{3} \cdot U_C \cdot I_L \cdot sen\varphi \quad (8.16)$$

$$S = \sqrt{3} \cdot U_C \cdot I_L \quad (8.17)$$

8.3 Delta-Balanced Load

When we have delta-connected loads (Fig. 8.3), they undergo both of the respective compound voltages. The current flows through every load.

We call I_{12}, I_{23}, e I_{31}, phase currents (I_F). Since this is a balanced system, we can write:

$$Z_{12}\angle\varphi_{12} = Z_{23}\angle\varphi_{23} = Z_{31}\angle\varphi_{31}$$

Recalling Ohm's law for a balanced system we have:

$$I_{\vec{12}} = \frac{U_{\vec{12}}}{Z_{12}\angle\varphi_{12}} \quad (8.18)$$

$$I_{\vec{23}} = \frac{U_{\vec{23}}}{Z_{23}\angle\varphi_{23}} \quad (8.19)$$

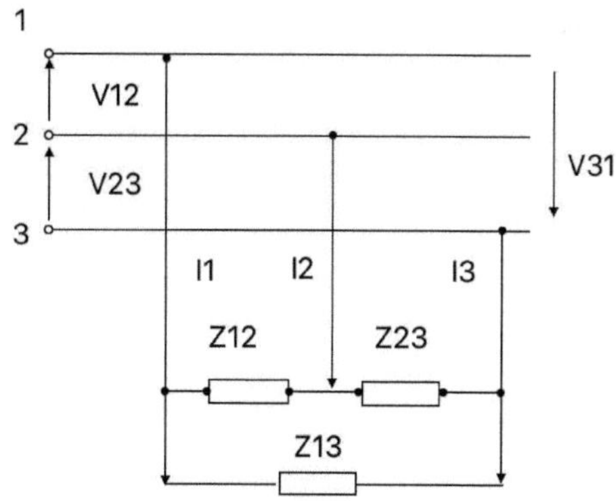

Fig. 8.3 Balanced load in delta. *Source* own elaboration

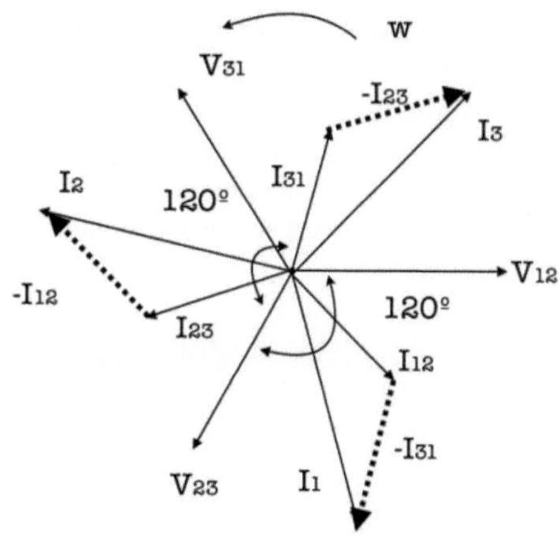

Fig. 8.4 Phasor diagram with compound voltages. *Source* own elaboration

$$\vec{I_{31}} = \frac{\vec{U_{31}}}{Z_{31}\angle\varphi_{31}} \tag{8.20}$$

Keep in mind that because the compound voltages are 120° out of phase with one another, the currents will be at an angle of φ to each of the corresponding compound voltages and will also be 120° out of phase with one another. Since the voltage that we apply and the impedances have the same modular value, the currents will likewise be the same.

$$|I\_12| = |I\_23| = |I\_31| = I\_f \tag{8.21}$$

Then, three additional currents, I_1, I_2 and I_3 appear in the supply line to the loads. We call these currents as line currents (I_L). For every node that appear in the delta load connections, we apply Kirchhoff's first to determine the correlation between these line currents and phase currents.

Node 1:

$$\vec{I_1} = \vec{I_{12}} - \vec{I}_{31} \tag{8.22}$$

Node 2:

$$\vec{I_2} = \vec{I_{23}} - \vec{I}_{12} \tag{8.23}$$

Node 3:

$$\vec{I_3} = \vec{I_{31}} - \vec{I}_{23} \tag{8.24}$$

Now, we draw a vector diagram with the compound voltages set at 120º intervals from each other (Fig. 8.4).

Assuming that the loads are inductive, we draw the phase currents lagging each of their respective compound voltages by an angle φ. If we draw the line currents by performing vector sum operations of their corresponding phase currents (refer to the relationships obtained above) we obtain:

Node 2:

$$\vec{I_2} = \vec{I_{23}} - \vec{I}_{12} \tag{8.25}$$

Node 3:

$$\vec{I_3} = \vec{I_{31}} - \vec{I}_{23} \tag{8.26}$$

Then, we have the line current vector diagram (Fig. 8.5). It shows the sum of $-I_{12}$ and $-I_{21}$.

We see that there exists a 30º angle between the phase and line currents.

An isosceles triangle appears where it is satisfied that:

$$\alpha = \beta \tag{8.27}$$

Using simple trigonometric, we have that:

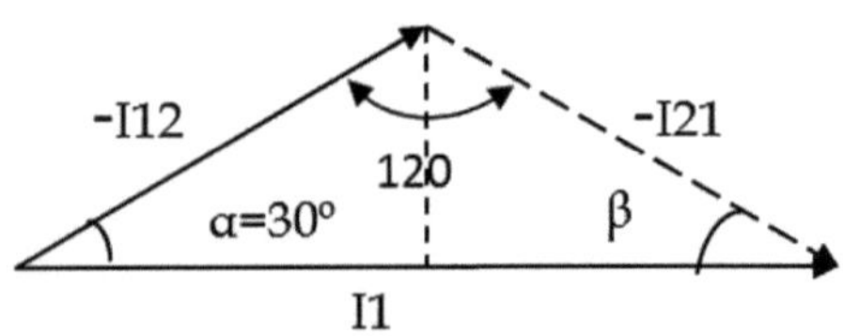

Fig. 8.5 Vector diagram with line currents. *Source* own elaboration

$$180^{\underline{o}} = \alpha + \beta + 120^{\underline{o}} \tag{8.28}$$

This implies:

$$\alpha = 30^{\underline{o}} \tag{8.29}$$

Only by applying basic trigonometry we determine the relationship between phase currents and line currents (Fig. 8.5).

$$cos\,30^\circ = \frac{I_1/2}{I_{12}} \rightarrow I_1 = 2 \cdot I_{12} \cdot cos\,30^\circ \tag{8.30}$$

$$I_1 = 2 \cdot I_{12} \cdot \frac{\sqrt{3}}{2} = I_{12} \cdot \sqrt{3} \tag{8.31}$$

Then, we conclude that:

$$I_L = \sqrt{3} \cdot I_f \tag{8.32}$$

The line currents have the same modulus value and lag 120° with one another. They are $\sqrt{3}$ higher than the phase currents.

8.3.1 Power of the Three-Phase System

Only by adding the power from each of the three separate single-phase loads we get the power that the delta-configured load produces, as we express before (Eq. 8.33).

$$P = U_{12} \cdot I_{12} \cdot cos\,\varphi_{12} + U_{23} \cdot I_{23} \cdot cos\,\varphi_{23} + U_{31} \cdot I_{31} \cdot cos\,\varphi_{31} \tag{8.33}$$

We know that the phase, line, composed voltages, as well as the power factors, are all equal in a balanced system. Then, we can write:

$$P = 3 \cdot U_C \cdot I_f \cdot cos\varphi \tag{8.34}$$

As we know:

$$I_f = \frac{I_L}{\sqrt{3}} \tag{8.35}$$

Substituting the phase current, we have that:

$$P = 3 \cdot U_C \cdot \frac{I_L}{\sqrt{3}} \cdot cos\varphi \tag{8.36}$$

From which it follows that:

$$P = \sqrt{3} \cdot U_C \cdot I_L \cdot cos\varphi \tag{8.37}$$

where P is the active power of the three-phase load, U_C is the composed voltage, I_L is the current line, and cos φ represents the power factor of the load.

We apply the same process to find the reactive power, given that:

$$Q = \sqrt{3} \cdot U_C \cdot I_L \cdot sen\varphi \tag{8.38}$$

$$S = \sqrt{3} \cdot U_C \cdot I_L \tag{8.39}$$

We conclude that, the equations for determining the output of a symmetrical three-phase mechanism hold true when applied to both star and delta-connected loads.

8.4 Electric Motor Efficiency

This section reviews the efficiency of the connected loads.

Question How much electrical power does the motor draw from the mains?

The electrical power consumption of the motor from the mains may not be clear. In this scenario, how much electricity the motor consumes from the grid? Nevertheless, we know the useful power that it can provide through its shaft and the efficiency at which it accomplishes this. For example, 5 horsepower at 85 % efficiency (η=85%).

We refer to P as the active electrical power that the motor consumes from the electrical network, and P_u as the useful or mechanical power the motor generates on its axis. The difference between these powers yields the power lost in the motor, Pp. Then, we have:

$$P_p = P - P_u \tag{8.40}$$

On the one hand, power losses due the Joule effect when current flows through the windings were defined as:

$$P = R \cdot I^2 \tag{8.41}$$

We also know that other losses like hysteresis and eddy currents in magnetic circuits, as well as friction in the bearings and ventilation, are the two main losses. The motor's efficiency denotes the percentage relationship between its useful power and overall or electrical power, i.e.:

$$\eta = \frac{P_u}{P} \cdot 100 \tag{8.42}$$

In our example, the motor's electrical power consumption from the grid is:

$$\eta = \frac{P_u}{P} = \frac{5 \cdot 736}{85} \cdot 100 = 4239W \tag{8.43}$$

8.5 Summary

This chapter focuses on the connection of receivers (loads) in three-phase electrical systems. It details the two primary connection methods: the star (Y) configuration and the delta (Δ) configuration for balanced loads. For each configuration, the chapter explains how to calculate the power of the three-phase system.

Finally, it introduces the concept of electric motor efficiency, which is a key measure of performance for a common type of three-phase receiver, relating its mechanical power output to its electrical power input.

Chapter 9
Asynchronous Machines

Abstract This chapter reviews the asynchronous machine. The versatility of this rotating electric machine makes them suitable for working as motors, generators, brake or transformers. The following sections present the wound rotor and the squirrel cage rotor. It is also important to review the constituent parts of the asynchronous machine. In order to analyze how asynchronous machines work, this chapter ends by analyzing their characteristic working curves.

Keywords Asynchronous machine · Excitation current · Electrical power system concepts · Squirrel cage rotor · Wound rotor · Constituent parts · Generator · Motor and transformer

9.1 Introduction

Question What is the most characteristic feature of asynchronous machines?

It is common to classify asynchronous alternating current machines according to the difference between the rotation speed and the synchronous speed. This means they can never reach synchronous speed, regardless of how they operate. Another term for this class of machine is "induction machines".

We use the term 'asynchronous machine' because they do not require an external excitation system, unlike other types of machines such as synchronous machines. The excitation current in this machine results from the emf that the interaction within the armature and the inductor causes. The cross-section of an induction machine shows the armature, the rotor, and the stator, as well as other elements that we present (Fig. 9.1).

D. Borge-Diez et al., *Fundamentals of Electrical Power Engineering*, Lecture Notes in Electrical Engineering 1540, https://doi.org/10.1007/978-3-032-18904-2_9

Fig. 9.1 Asynchronous machine. *Source* from wikicommons under free to use license [1]

9.2 The Electric Power System. Concepts

The following is an overview of a number of concepts that arise in the analysis of asynchronous or induction machines. The stator frequency denotes the frequency of the network powering the stator winding, and the frequency of the current running through this winding. Knowing the stator frequency is valuable because, together with the number of pole pairs in the machine, it determines the value of the synchronous speed. The synchronous speed equates the speed of the rotating field that the inductor creates.

We determine the synchronous speed using one of the following equations:

$$n_1 = \frac{f_1}{p}\begin{cases} n_1 = rps \\ f_1 = Hz \\ p = \text{number of pole pairs} \end{cases} \tag{9.1}$$

Or,

$$n_1 = \frac{60 \cdot f_1}{p}\begin{cases} n_1 = rps \\ f_1 = Hz \\ p = \text{number of pole pairs} \end{cases} \tag{9.2}$$

The rotational speed refers to the speed of the rotor. In asynchronous machines, it is true that:

$$n_2 \neq n_1$$

By convention, we give a positive sign to the synchronous speed.

For this reason, if the rotor rotates in the same direction as the rotating field, its speed is positive; conversely, if the rotor rotates in the opposite direction, its speed is negative.

Now, we define slip s as:

$s = \frac{n_1 - n_2}{n_1}$, where $n_s = n_1 - n_2$ is referred to as slip speed.

Rotor frequency (fs) denotes the voltage and current frequency that appear in the rotor windings. The relative speed of the rotating field and the rotor determines its value. We define it as:

$$f_s = p \cdot (n_1 - n_2) = p \cdot n_s \tag{9.3}$$

where n_1, n_2 and n_s are given in revolutions per second (rps) and frequency in hertz (Hz).

Remember that 1 rps = 1 Hz.

Therefore, it follows that:

$$f_s = p \cdot s \cdot n_1 \rightarrow f_s = s \cdot f_1$$

9.3 Classification of Asynchronous Machines

Apart from the various applications they have, we classify asynchronous machines according to the following features.

9.3.1 According to the Number of Phases

The number of phases refers to the stator winding, which corresponds to the number of phases of the grid it connects to. Accordingly, they are categorized as follows:

- Single-phase.
- Polyphase (three-phase machines are the most commonly used).

9.3.2 According to the Number of Poles

According to number of poles the machine has, we classify them as two-pole, three-pole, four-pole machine, and so on. The number of poles influences the synchronous speed of the machine. As a result, the range of speeds that the asynchronous machine can achieve determines which type to choose.

9.3.3 According to the Type of Rotor

Depending on the constructive features of the rotor, asynchronous machines take on various forms:

- Wound rotor.
- Squirrel cage.

9.3.3.1 Wound Rotor

In the wound rotor case, we place the wires on open or half-open slots distributed along the outer periphery of the magnetic plate cylinder that forms the rotor. This builds the electric circuit of the machine. The scheme of a wound rotor electric machine can be differently performed (Fig. 9.2).

For example, a scheme of an electric machine the three phases may connect to the rotor with slip rings (Fig. 9.3).

It is a three-phase circuit with terminals that connect to slip rings in a star configuration. Slip rings often locate on the shaft. We can also find a star-connected rotor circuit scheme (Fig. 9.4).

These rings provide access to the rotor circuit via stationary brushes that contact them.

Question Why do we want access to the rotor circuit via brushes?

Fig. 9.2 Squirrel cage rotor cross-section of an electric machine. *Source* from wikicommons under free to use license [2]

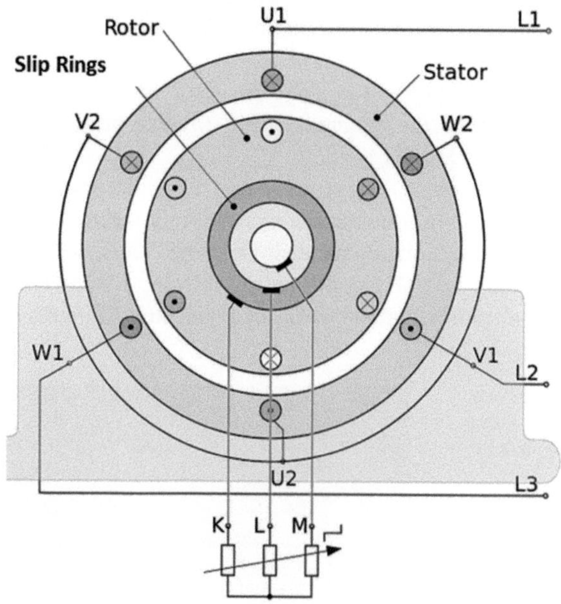

Fig. 9.3 Wound rotor. *Source* from wikicommons under free to use license. *Source* from wikicommons under free to use license [3]

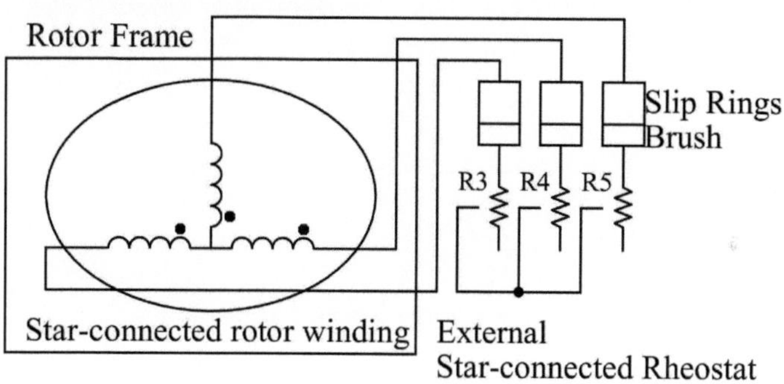

Fig. 9.4 Slip ring three-phase induction motor schematic. *Source* own elaboration

In order to answer this question, our aim is to change the rotor circuit's traits by adding resistors, among other things. We will go over the reasons for this below.

9.3.3.2 Squirrel Cage Rotor

The conductor bars, which are usually made of copper or aluminum, are what make up the electrical circuit of a squirrel cage rotor. These bars are positioned in slots, which are typically closed, around the magnetic package of the rotor. In a squirrel cage rotor of an induction motor the magnetic package containing the conductor bars is visible (Fig. 9.5).

The main components of a squirrel-cage rotor including the shaft (1), the conductor bars (2) as well the magnetic plates (3) that house the conductor bars and the rotor (4) as well (Fig. 9.6).

The advantage of the squirrel cage rotor is that it is much stronger than the wound rotor. Unfortunately, a major drawback is that the rotor electrical circuit is inaccessible, making any attempt at maintenance impossible. Thus, connecting resistors to the rotor winding is impractical. It is interesting showing a squirrel cage rotor with a 90° cut in its magnetic plates, and we see the cage (electrical circuit) inside (Fig. 9.7).

Fig. 9.5 Squirrel cage rotor of an induction motor. From wikicommons under free to use license [4]

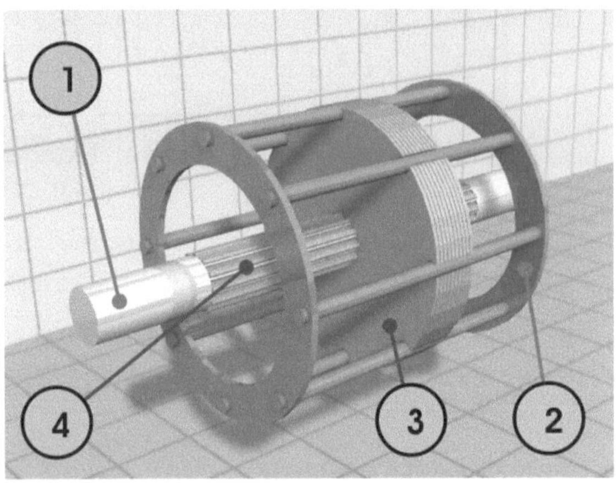

Fig. 9.6 Scheme of a squirrel cage rotor induction motor. *Source* from wikicommons under free to use license [5]

Fig. 9.7 Motor with a squirrel cage induction motor. *Source* from wikicommons under free to use license [1]

9.4 Constituent Parts and Their Function

Question What are the basic constituents of an asynchronous machine?

Like all rotating machine, asynchronous machines include stator, rotor and air gap. The stator winding acts as the inductor and comprises a bundle of magnetic plates with slots on its inner cylindrical contour (the closest to the rotor), regardless of the type of rotor it contains.

As was previously mentioned, the coils that comprise the stator winding are dispersed throughout the slots, for example in a three-phase stator machine (Fig. 9.8).

The armature comprises the rotor winding, which can be a wound rotor or a squirrel cage type, as showed in the previous section. The wound rotor must have the same number of poles as the stator. For a squirrel cage rotor, the stator's field creates an equal number of poles in both the rotor and stator. The air gap is the distance

Fig. 9.8 Stator of an electric machine with windings. *Source* from wikicommons under free to use license [6]

Fig. 9.9 Squirrel cage rotor with fans. *Source* from wikicommons under free to use license [7]

between the two and, along with the stator and rotor magnetic packages, completes the machine's magnetic circuit.

The cooling system and the housing are essential auxiliary components. To dissipate the heat that the machine generates during its operation, the cooling system employs the rotation of the shaft to facilitate air circulation through embedded fans. The cooling system uses the shaft's rotation to promote airflow through fans embedded in order to dissipate the heat produced by the apparatus's losses, lowering the efficiency of the machine (Fig. 9.9).

Question How can we improve heat dissipation?

Axial and radial ventilation channels in the stator and rotor help dissipate heat. It also provides a way to install both passive and active cooling systems. An induction machine's housing serves two main purposes. First of all, it provides mechanical protection against external threats. The case protects, and holds some of the electric components that comprise the system (Fig. 9.10). It includes the stator, rotor, cover and two sets of bearings with their grease seals.

Second, it is a safety feature for people. It prevents them from coming into contact with potentially hazardous components (because they are live or rotate at high speed).

9.5 Operating Modes

Some of the most interesting modes this machine operates are those that relate to the motor, the generator, the transformer, and the brake.

Fig. 9.10 Servicing the AD5 asynchronous motor of the Yauza-209–1 tape recorder. *Source* from wikicommons under free to use license [8]

9.5.1 *Motor*

This is the most common application of the induction motor. In this mode, the stator winding connects to an electrical supply (single-phase or three-phase, depending on the motor type), and the rotor winding must be short-circuited.

Recalling that:

n_1 is the synchronism speed of the magnetic field, n_2 is the rotor speed, and s is slip.

For the asynchronous machine working as a motor, the rotating field speed is:

$$0 \rightarrow < n; n_2 < n_1 \leftrightarrow 1 \leftrightarrow 1 > s > 0$$

9.5.2 *Generators*

We use the asynchronous generator in small power plants or equivalent cogeneration facilities that demand the rotor to operate at speeds higher than the synchronous speed. The rotor winding must be short-circuited, which leads to the asynchronous generator providing active power to the AC mains linked to its stator. Therefore, the asynchronous generator provides active power to the AC mains linked to its stator, which subsequently distributes power to the corresponding loads.

Some questions that arise are:

Question What is the impact of adding capacitors?

The addition of capacitors boosts the power, providing the reactive power that the machine requires for its magnetization.

Question What components is induction generators connected to?

Generally, asynchronous generators connect to grids and work together with the synchronous generators of major power stations.

Question Why synchronous generators maintain asynchronous generators?

This is because synchronous generators determine both the voltage and the frequency of the network's operation.

9.5.3 Transformers

Induction machines can work as transformers, provided that their rotor and stator electrical circuits are accessible. However, this mode of operation is not viable for squirrel-cage rotor induction machines. To operate as a transformer, the rotor must be locked, resulting in a fixed speed and slip. We can draw the windings in an asynchronous machine (Fig. 9.11). The rotor condition means that slip is unit. Then, we have:

$$n_2 = 0 \leftrightarrow s = 1$$

Under these circumstances, if one winding (usually the stator winding) connects to an active electrical network, in the other winding (usually the rotor winding) appear a set of voltages of proportionate value (depending on the number of turns of each winding) and equal frequency. In effect, the windings serve as the primary and secondary functions of a transformer. Finally we have the equivalent circuit of an asynchronous machine (Fig. 9.12).

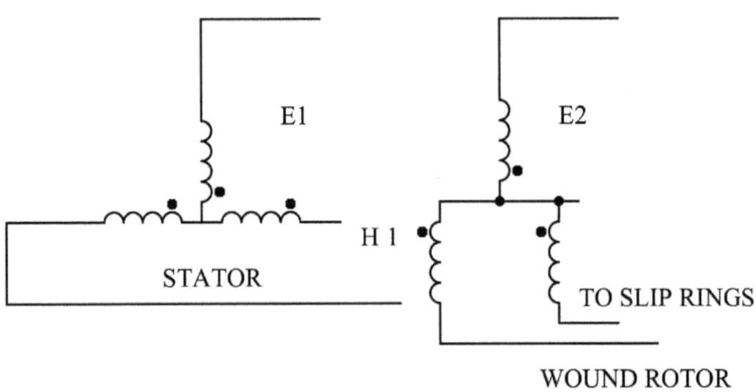

Fig. 9.11 Scheme of an asynchronous machine acting as a transformer. *Source* own elaboration

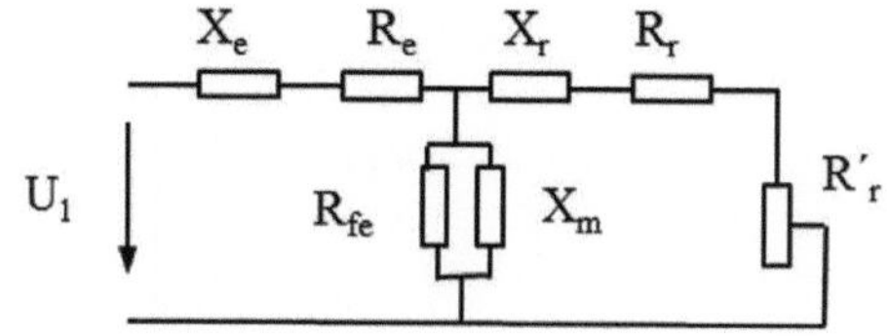

Fig. 9.12 Equivalent circuit of an asynchronous machine. *Source* own elaboration

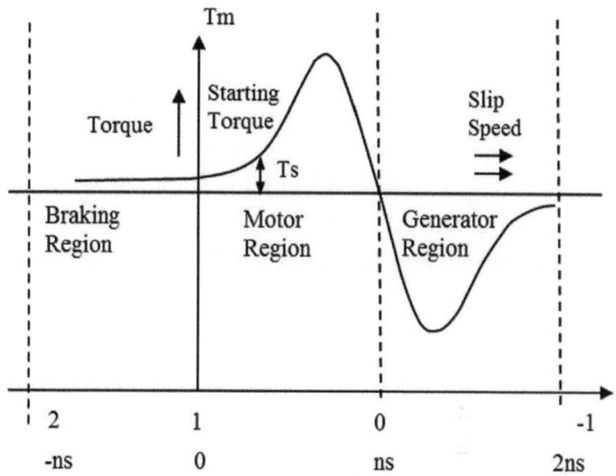

Fig. 9.13 Asynchronous machine operation diagram. *Source* own elaboration

This application has the advantage that, depending on where the rotor is fixed with respect to the stator, it is possible to achieve any phase difference angle between the primary and secondary voltages. For example, in electrical laboratories, it is necessary to adjust the angle of phase difference between primary and secondary in order to test meter and wattmeter. Now we plot the operational modes of the asynchronous machine that provide a torque-slip speed graphical representation (Fig. 9.13).

9.5.4 Brake

When the asynchronous machine acts as a brake the shaft of the machine must rotate against the direction of the rotating field. We can plot the behavior of the asynchronous machine acting as a brake (Fig. 9.14). We express it as:

$$n_2\langle 0 \leftrightarrow s\rangle 1$$

In this mode of operation, appears a torque that opposes the torque applied to the shaft, provoking that the machine stops which can be achieved with variable resistors (Fig. 9.15). Including variable resistors limit the currents to the loops.

In this mode of operation, the rotor winding must be short- circuited, while the stator winding must connect to an active electrical network.

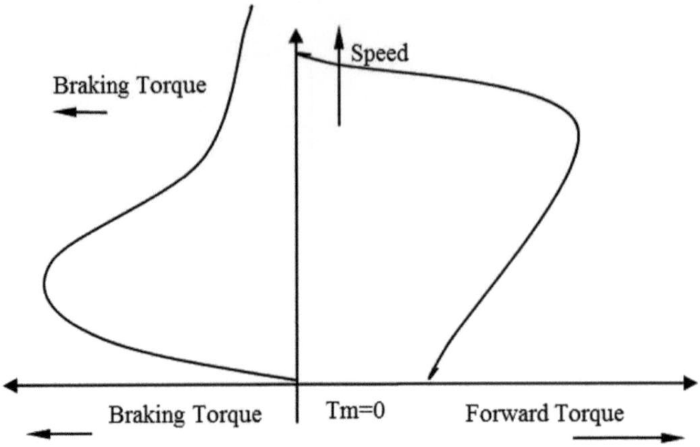

Fig. 9.14 Rheostatic braking characteristics of a 3-phase induction machine. *Source* own elaboration

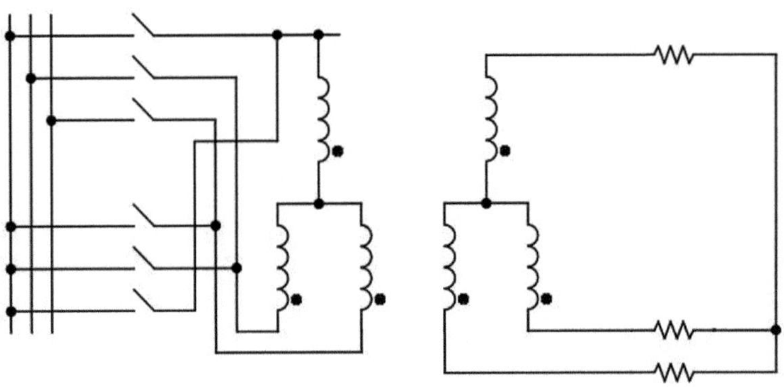

Fig. 9.15 Connections for plugging of a three-phase induction motor with rotor resistance for current limiting and braking torque. *Source* own elaboration

9.6 Rotating Field in Asynchronous Machines

9.6.1 Arago's Experiment

The Arago's disk contains a conductive disk that can rotate around its rotation axis (Fig. 9.16a). Beneath it, locates a U-shaped magnet that is also able to rotate around the same axis. With no disk magnetization, we see how the magnet's rotation causes the disk to rotate. When the magnet turns, the copper disk above it (suspended by a string) rotates. Rotating a copper disk within a magnetic field generates a current through electro-magnetic induction. Faraday's dynamo (1832) intends to explain Arago's disk by the phenomenon of electromagnetic induction (Fig. 9.16b).

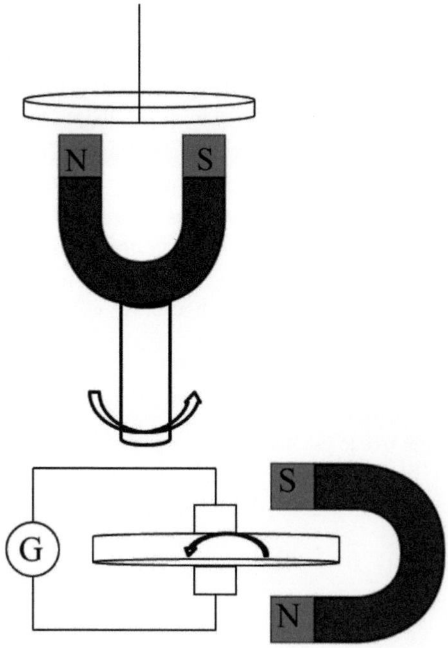

Fig. 9.16 **a** and **b** Arago's disk and Faraday's dynamo. *Source* own elaboration

We assume that the magnet rotates counterclockwise. Either way, a conductive material receives electromotive forces that are induced on the disc close to each pole, as per Faraday's law. This generates currents that circulate through the metal disk. The magnetic field that the magnet creates influences the metal disk. According to Faraday's law, we plot the emf effects (Fig. 9.17).

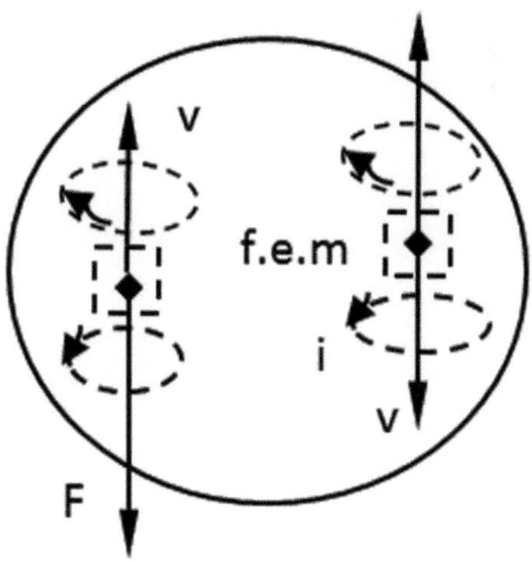

Fig. 9.17 Electromotive force. *Source* own elaboration

Laplace's law states that two forces are thus produced, causing the disk to rotate counterclockwise like the magnet, following the right-hand rule of forces (Fig. 9.18).

This creates a pair of forces that cause the disk to rotate counterclockwise like the magnet, in accordance with Laplace's law. It is important to highlight the speed of the copper disk as well of the magnet (Fig. 9.19).

The rotation speed of the disk remains lower than the speed of the magnet. Note that if the speeds are equal, no relative speed with respect to the magnet exists, thus there is no electromotive force and the process would not start. Therefore, it is necessary relative speed between the magnet and the disk for the electromotive forces and currents to appear. It's worth saying that the effect we obtain is different to that

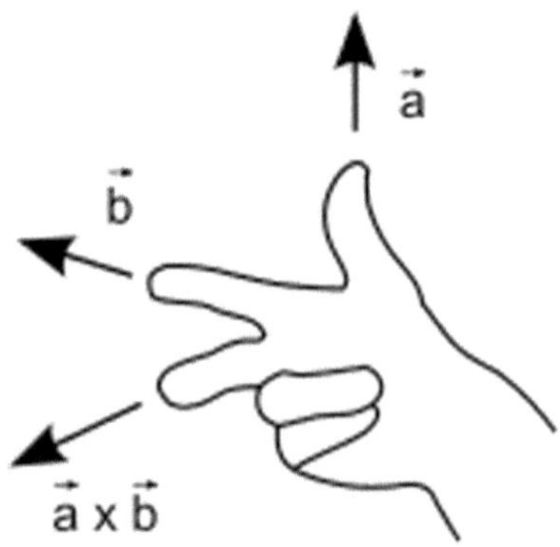

Fig. 9.18 Right-hand rule. *Source* from wikicommons under free to use license [9]

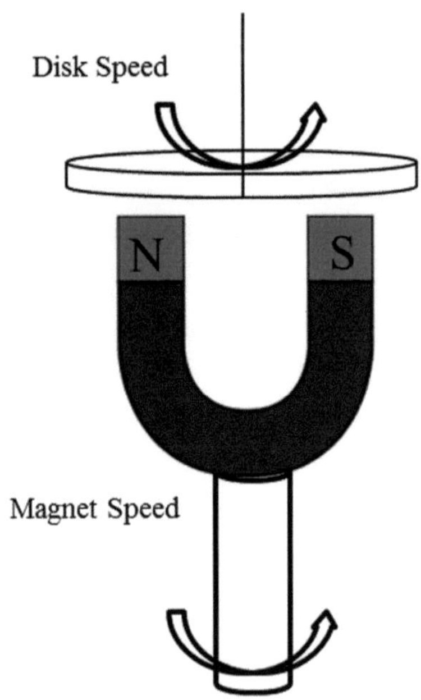

Fig. 9.19 Velocities of magnet and copper disk. *Source* own elaboration

of an electric motor, since the mechanical energy that the magnet's rotation supplies transforms into mechanical energy in the disc's rotation, as opposed to electrical energy transforming into mechanical energy. However, this example demonstrates the application of the fundamental laws that validate the operating principle of the induction motor.

9.6.2 Ferraris Theorem

The Ferraris theorem states that a polyphase winding, when traversed by a polyphase system of three-phase currents, generates a rotating field that rotates at synchronous speed. The frequency of the three-phase currents passing through the windings determines this speed, as stated in the following theorem. We see the rotating magnetic field that generates a 3-phase input signal (Fig. 9.20). We define:

$$p \cdot n = 60 \cdot f \,(\text{n in rpm}) \tag{9.4}$$

This theorem applies to both stator and rotor windings.

9.7 Three-Phase Motor Operation

Note how the coils place carefully in the stator slots. As we see, the stator's winding comprises a collection of these slots that host the designed loops (Fig. 9.21).

Question Where are the three phase windings?

The stator of a three-phase induction motor contains the windings for the three phases. We illustrate how to connect the stator windings of an asynchronous machine in either star or delta configuration (Fig. 9.22).

9.8 Asynchronous Motor Operation

The stator creates a three-phase current system with i_1 (t), i_2 (t), and i_3 (t) flowing through its windings when its three phases receive a balanced three-phase voltage system (as shown in the right figure above). According to Ferraris' theorem, this current system produces a rotating field at a constant speed. We can plot the rotation of the magnetic field that the stator phases generate (Fig. 9.23). The rotor follows the rotating magnetic field but rotates asynchronously, at a lower speed.

According to Faraday's and Lenz's law, it appears magnetic flux due to induced rotor current (Fig. 9.24).

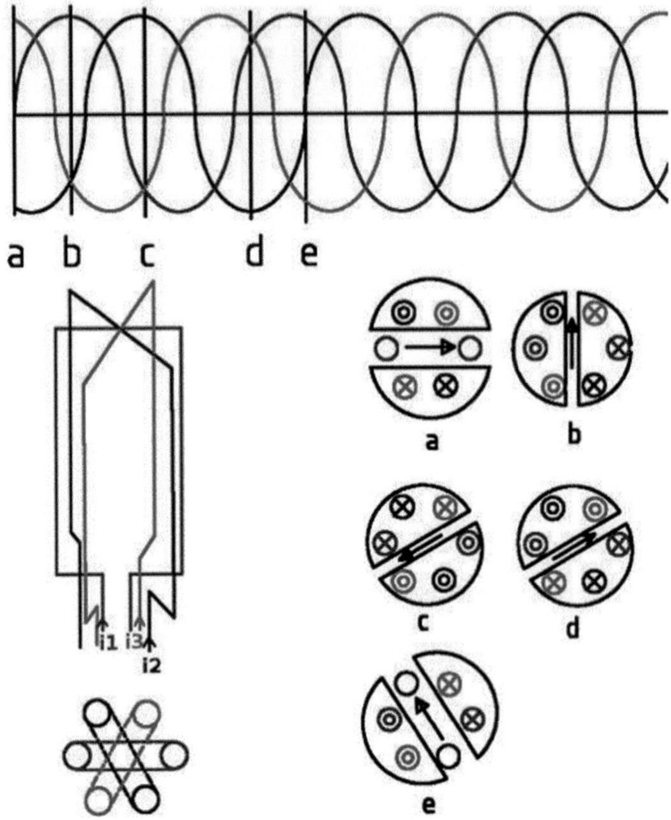

Fig. 9.20 Rotating magnetic field. *Source* from wikicommons under free to use license [10]

Fig. 9.21 Stator with slots that house windings. *Source* from wikicommons under free to use license [11]

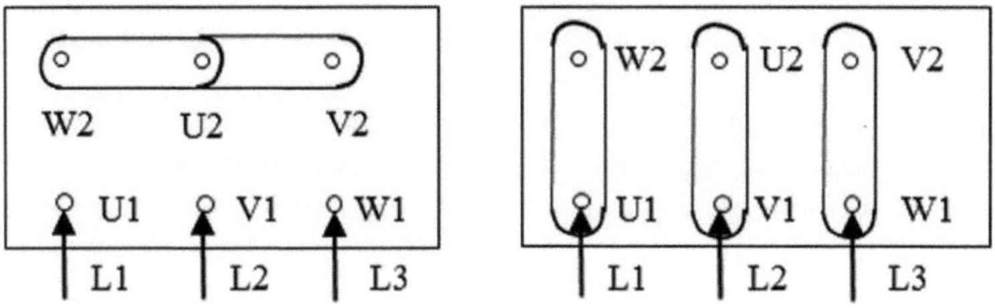

Fig. 9.22 Three-phase induction motor's stator windings wiring. Star (left), and delta (right) connection. *Source* own elaboration

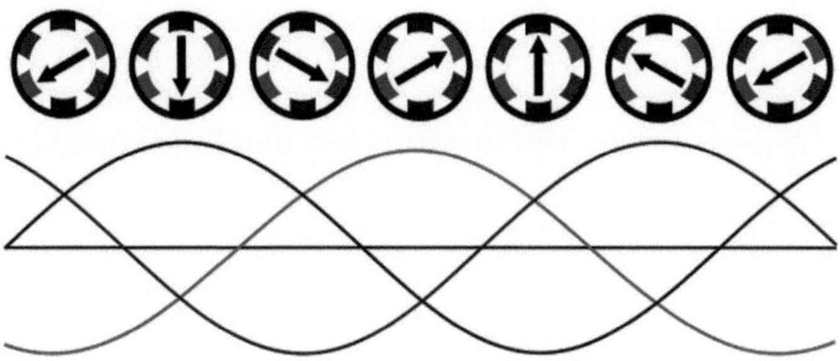

Fig. 9.23 Three phase voltage induction waveform. *Source* from wikicommons under free to use license [12]

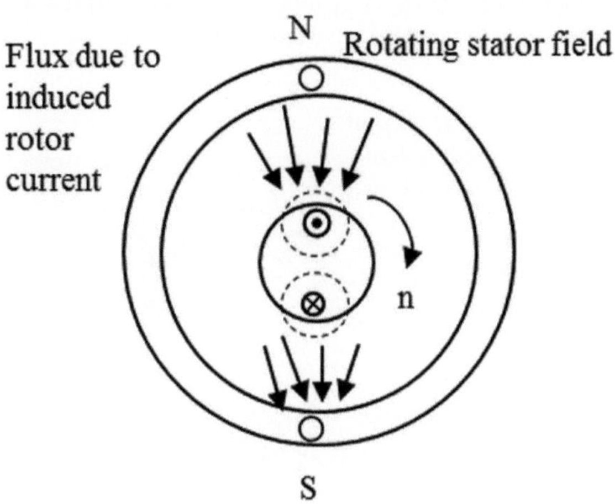

Fig. 9.24 Flux due to induced rotor current. *Source* own elaboration

The rotor windings experience electromotive forces because of this rotating field. Current flows through the rotor windings as a result of the connections made between them, either by short-circuiting the windings with the rings in the squirrel cage, or by connecting the ends and beginnings. These groups of currents, under the influence of a magnetic field, generate forces and torques according to Laplace's law. Below we see the forces and components over a squirrel-cage rotor (Fig. 9.25).

These forces and torques system rotate the rotor at a speed n_2 in similar direction as n_1. However, it is important to note that the final rotor speed, n_2, never equals n_1 due to Faraday's law, which requires a relative speed between the magnetic field and conductor (Fig. 9.26).

Keep in mind that the rotor and magnetic flux speed are very close, but they are not similar. When the rotor rotates at n_2, the slip velocity (n_S) between the rotor and the stator rotating field causes the frequency of the rotor's electromotive forces as well as the frequency of the rotor currents to occur. This frequency is expressed as:

$$f_s = \frac{p \cdot n_s}{60} = \frac{p \cdot s \cdot n_1}{60} = s \cdot f_1 \tag{9.5}$$

In a vacuum, meaning that no load connects the rotor shaft, the rotor speed is almost the same of the magnetic field. Consequently, n_S is almost zero. When the machine loaded, the speed increases but it is still close to zero. A low n_S value suggests that the frequency of rotor currents is low, lowering the iron losses in the machine's rotor. Therefore, the sheet of metal that we use to manufacture rotors is of lower quality than the sheet metal we use to make stators.

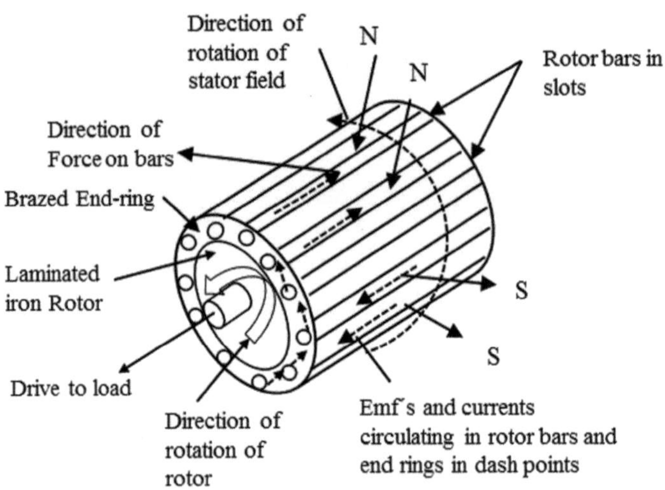

Fig. 9.25 Direction of forces in a squirrel-cage induction machine. *Source* own elaboration

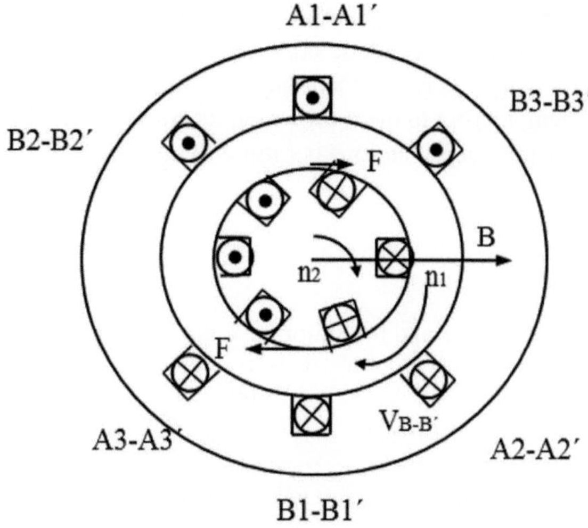

Fig. 9.26 Relative speed between rotor and magnetic flux. *Source* own elaboration

9.8.1 Final Rotating Field Velocity

In the asynchronous motor, regardless of the rotor speed, which can vary, the frequency regulation mechanism f_S in the rotor phases facilitate the "coupling" of the stator and rotor magnetic fields. The stator and rotor fields couple at synchronous speed, even if the rotor rotates at a dissimilar speed. This happens when the stator and the rotor have the same number of poles (Fig. 9.27).

As we already know the rotor produces electromotive forces with frequency that depends on the relative speed between the rotating field of the stator (which rotates at n_1) and the rotor (which rotates at n_2).

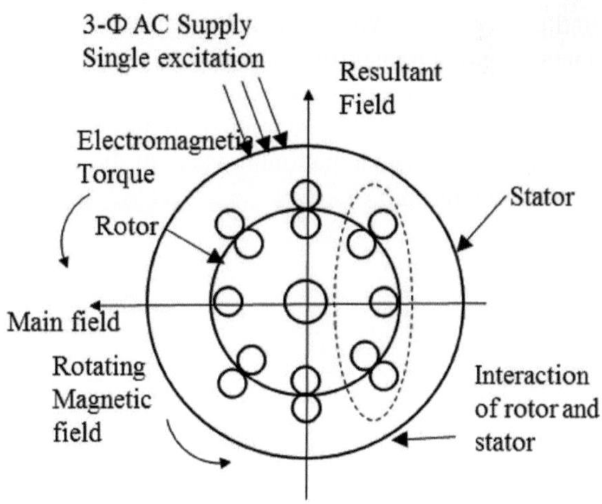

Fig. 9.27 Three-phase induction motor interactions. *Source* own elaboration

The electromotive forces cause three-phase currents to flow within the rotor at a frequency of f_S. These currents create a rotating magnetic field within the rotor, according to Ferrari's theorem. This field has a speed that depends on the frequency f_S, with respect to the rotor's motion. As a result, the rotating field rotates at a speed (measured in revolutions per minute) that is:

$$\frac{(60 - fs)}{p_{\text{rotor}}}$$

Since the rotor and the stator have the same number of pole pairs, we can write:

$$\frac{60 \cdot f_s}{p_{\text{rotor}}} = \frac{60 \cdot f_s}{p} = n_s$$

As a result, the rotating field of the rotor will rotate at a speed $n_2 + n_S$ with respect to an observer in the stator because the rotor is not fixed and rotates at a speed n_2.

Therefore, it follows that:

$$n_S + n_2 = n_1 - n_2 + n_2 = n_1$$

As a result, and as we see Fig. 9.27, when the number of poles of the stator and rotor coincide, the revolving magnetic fields that the rotor and stator generate rotate at the synchronous speed (n_1), regardless of the rotor's rotational speed (n_2).

9.9 Comparison Between Motor and Transformer

This section looks at how interact the voltage and the current that stator and rotor windings generate. We make it by means of drawing comparisons between the functioning of a transformer and of an asynchronous motor.

Suppose we have an asynchronous motor that generates a rotating field at synchronous speed n_1 whose rotor spins at n_2 and a sinusoidal mutual flux (between stator and rotor) with a maximum value of Φ_0.

With this, the slip is:

$$s = \frac{n_1 - n_2}{n_1}, \tag{9.6}$$

and slip frequency is then:

$$f_s = \frac{p \cdot n_s}{60} = \frac{p \cdot s \cdot n_1}{60} = s \cdot f_1 \tag{9.7}$$

Similar to what happens in transformers, due to the resulting rotating field, an emf value also generates an emf wave.

$E_1 = 4,44 \cdot N_1 \cdot \Phi_0 \cdot f_1$, where N_1 is the number of turns per phase in series in the stator, while f1 corresponds to the frequency of the stator supply network.

The rotor also experiences a corresponding phase voltage in response to this same mutual interaction.

$$\mathrm{E_S = 4,44\ N_2 \cdot \Phi_0 \cdot f_S = 4,44\ N_2 \cdot \Phi_0 \cdot s \cdot f_1} \tag{9.8}$$

where N_1 represents the number of series loops per phase of the stator and f_1 denotes the frequency of the stator supply network. In the case of the rotor being still (as it happens at start-up), the rotor speed equals zero ($n_2 = 0$), the slip ratio equals unity ($s = 1$), and hence the rotor frequency is equivalent to the stator supply frequency ($f_S = f_1$).

In this case, the induced electromotive forces per phase on the rotor hold a specific value of:

$$E_S(s = 1) = E_2 = 4,44 \cdot N_2 \cdot \Phi_0 \cdot f_1 \tag{9.9}$$

Suppose the rotor constant speed, then:

Question What are the rotor speed n_2, the slip s, and the rotor frequency fs?

The emf that appears on the rotor in this case is zero.

$$E_S(s = 1) = E_2 = 4,44 \cdot N_2 \cdot \Phi_0 \cdot f_1 \tag{9.10}$$

Therefore, in any rotor rotation scenario, it is true that:

$$E_S = s \cdot E_2$$

Getting an asynchronous motor started is the same that an unplanned short-circuits in the transformer, resulting in high circulating currents during this process. Motor operation under load is similar to transformer operation under load. During no-load operation (when no mechanical load connects to the shaft, and it only has to overcome its own frictional torque), the rotor speed is very near to the synchronous speed. Therefore, the motor's no-load operation (with no mechanical load connected to the shaft) corresponds to the transformer's no-load operation (with no electrical load connected to the secondary).

9.10 Power and Torque in Three-Phase Asynchronous Motor

9.10.1 Power Distribution in the Asynchronous Motor

Question How does an asynchronous motor transfer power?

We know that an asynchronous motor converts electrical energy into mechanical energy. The magnetic flux transfers the energy from the stator to the rotor through the air gap. The process has losses that occur in different parts of the machine impacting the conversion process' efficiency.

The analysis that follows assesses the power balance in the operation of a three-phase asynchronous motor. Let V_1 represents the voltage per phase, I_1 the current per phase, and φ1 the phase difference between both vectors, the power that absorbs from the grid is:

$$P_1 = 3 \cdot V_1 \cdot I_1 \cdot cos\varphi_1 \tag{9.11}$$

Some of the power that enters the stator is then emitted as heat due to the Joule effect in its windings. This power becomes:

$$P_{J1} = 3 \cdot R_1 \cdot I_1^2 \tag{9.12}$$

Hysteresis and Foucault losses cause power losses in the stator (PFe_1). As long as these depend on the frequency, we assume that iron losses only occur in the stator because the frequency of the currents in the rotor windings are very low and the slip is typically small at the operating point (e.g., for $s = 5\%$ with $f_1 = 50$ Hz, it results in $f_S = 2.5$ Hz $< < f1$).

Question Then, we ask what the stator losses are.

P_{J1} and P_{Fe} represent the overall losses in the stator. Therefore, we can find P_{12}, or the power that the stator transmits to the rotor, which is the electromagnetic power that pass through the air gap and reach the rotor.

$$P_{12} = P_1 - P_{J1} - P_{Fe} \tag{9.13}$$

Question What are the losses due to the Joule effect in the rotor?

Additionally, the rotor suffers Joule losses (P_{J2}) with a specific value of Eq. 9.11.

$$P_{J2} = 3 \cdot R_2 \cdot I_2^2 \tag{9.14}$$

Because the iron losses in the rotor are so small, the power that reaches the shaft of the machine, and we call it internal mechanical power P_{mi}.

$$P_{mi} = P_{12} - P_{J2} \tag{9.15}$$

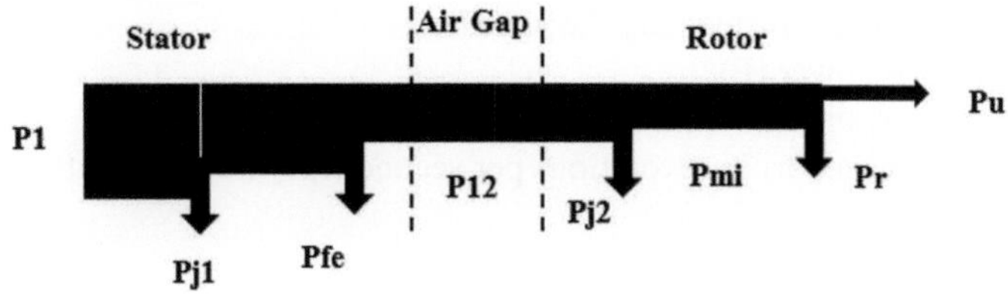

Fig. 9.28 Power balance in the motor. *Source* own elaboration

However, the useful power at the shaft is slightly lower due to friction and ventilation, provoking mechanical losses. Figure 9.28 depicts the power balance evolution in the motor. If we call mechanical losses as PR and useful mechanical power, as Pu, we write:

$$P_u = P_{mi} - P_R \tag{9.16}$$

9.10.2 Relationship Between Power and Losses. Efficiency

We use the ratio (Eq. 9.14) to determine the motor's efficiency since the output power comes from an input power P_1:

$$\eta = \frac{P_u}{P_1} = \frac{P_u}{P_u + P_R + P_{J2} + P_{Fe} + P_{J1}} \tag{9.17}$$

Similarly, the rotor's electrical efficiency is:

$$\eta_{rotor} = \frac{P_{mi}}{P_{12}} = 1 - s \tag{9.18}$$

Only by analyzing this expression we see that if the slip is low, the Joule losses in the rotor decrease and then the rotor efficiency increases. Therefore, it is helpful for these machines to operate with low slip.

9.11 Torque Study

This section reviews the torque that an asynchronous machine develops. Recalling the definitions of:

$Work = \tau \cdot \Delta\theta$

(Work;τ: torque $\cdot\Delta\theta$ change in angular displacement)

$K = \frac{1}{2} \cdot I \cdot w^2$ (K = Kinetic energy; I moment of inertia; w: angular velocity)

$P = \tau \cdot w$ (P power; torque; angular speed)
Remember that power (P): .
Now on we look on the torque calculation (Cm).
If we express n_1 and n_2 in revolutions per second (rps) we have that:

$$Cm = \frac{P_{mi}}{2\pi \cdot n_2} = \frac{P_{mi}}{2\pi \cdot n_1 \cdot (1-s)} \tag{9.19}$$

$$\text{Recalling that : s} = \frac{\mathrm{n}_1 - \mathrm{n}_2}{\mathrm{n}_1} \rightarrow n_2 = n_1(1-s) \tag{9.20}$$

We can also define torque through the power that the stator transfers to the rotor when they are rotating in sync. Substituting (Eq. 9.15) we have:

$$Cm = \frac{P_{mi}}{2\pi \cdot n_2} = \frac{P_{mi}}{2\pi \cdot n_1 \cdot (1-s)} = \frac{P_{12}}{2\pi \cdot n_1} \tag{9.21}$$

This statement of torque is true, as the denominator comprises the speed of the rotating field, which remains constant in the motor as long as the power supply frequency remains unchanged. For this reason, the power that the stator transmits to the rotor (P_{12}) refers to as the motor torque in synchronous watts.

9.11.1 *Discussion of the Motor Torque*

Question How does C_m vary as a function of n_2, R_2, V_1, y f_1?

After obtaining the torque expression, we examine how torque changes with rotor speed, rotor resistance, supply voltage, and mains frequency. Next, we will cover the characteristics curves of operation zones in asynchronous motors and the features that set apart various operating modes in asynchronous machines.

This section concludes with the analysis of the stable and unstable operation zones of the asynchronous motor, as well as the operating regimes of the asynchronous machine, through an outlook of the torque expression.

Torque change as a function of the rotor speed n_2.

Question What determines an asynchronous motor's operating speed?

The torque that an asynchronous motor develops and the opposing torque that the mechanical load imposes determine the speed at which the motor runs. According to the elastic torque-speed curve for a mechanical load can be of various types (Fig. 9.29).

First: constant torque is independent of speed, C_{rA}. This torque applies to loads that consume power in proportion to their speed, such as cranes and lifts.

Second: the relationship between torque and speed is linear, C_{rB}. This is the case of loads with power that depends on the square of the speed. For instance, we use

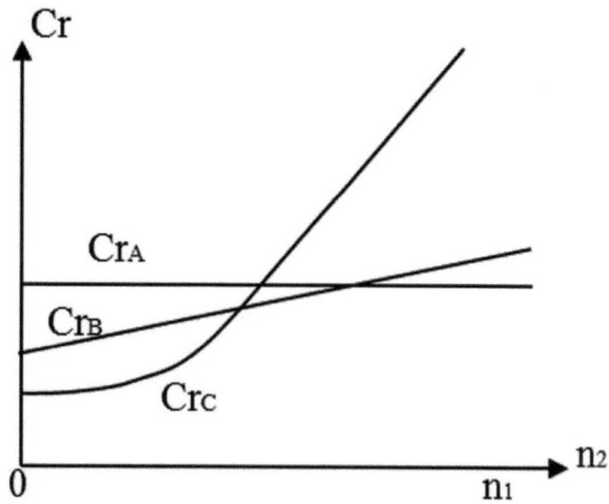

Fig. 9.29 Torque-speed curve types. *Source* own elaboration

them in machinery for the smoothing and production of satin finishes for materials, including fabric and paper.

Third: torque is a function of the square of the speed, C_{rC}. For example, we find it in loads whose power depends on the cube of the speed. For example, centrifugal pumps, fans and, more generally, loads associated with fluid motion.

Question What is the rated operating speed of a motor?

Running speed is the steady speed that a motor (n_2) reaches while driving a mechanical load after overcoming the start-up phase. We determine this number by using the motor-load system's motion (Eq. 9.17).

$$Cm - Cr = J \cdot \alpha = J \cdot \frac{d(2\pi \cdot n_2)}{dt} \tag{9.22}$$

where J is the moment of inertia of the motor-load assembly, in kg m^2, α is the angular acceleration in rad/ s^2, Cr is the resistive torque (including the torque of the mechanical load couples the shaft and that due to the mechanical losses of the motor), in Nm. Remember that 1 Newton = 1 kg m/s^2.

The resisting torque opposes the motor torque is what causes the negative sign (Eq. 9.17).

Question What is the angular acceleration the motor experiences when it reaches the rated operating speed?

When the asynchronous machine reaches the operating speed, the rotational speed is constant. Then, we can write:

$$\alpha = \frac{d(2\pi \cdot n_2)}{dt} = 0 \tag{9.23}$$

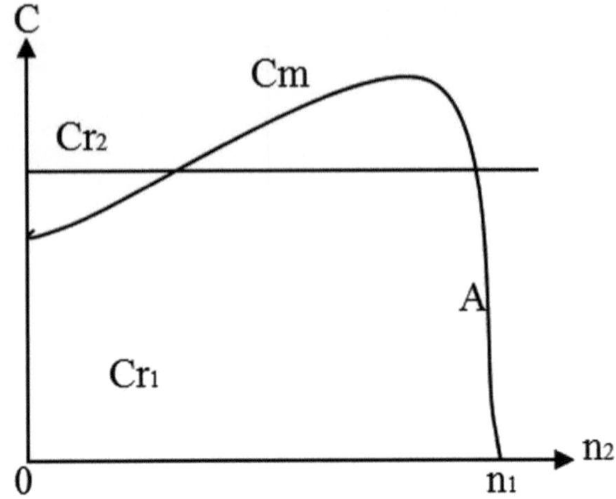

Fig. 9.30 Operating point. *Source* own elaboration

Therefore, at the operating point, the motor's torque equals the resisting torque (Fig. 9.30).

$$Cm = Cr$$

Plotting the torque and resistance torque curves together as a function of speed indicates that the operating speed is at the intersection of the two curves. This is, at the point where the two curves intersect (point A in Fig. 9.30 for the pairs C_m and C_{r1}).

Question For the system to work, what conditions need?

For the system to work, the motor torque at the start-up ($n_2 = 0$, $s = 1$) must be greater than the resisting torque. Thus, with the resisting torque C_{r2} Fig. 9.31 the motor would not start.

Question What are the normal slip values for a motor?

The typical slip values for a motor are approximately 0.1 when it operates at maximum speed.

Question What is the speed at which the rotor rotates?

Recalling the Equation of slip:

$$s = \frac{n_1 - n_2}{n_1}$$

This means that the rotor rotates very close to the synchronous speed.

Increasing the rotor resistance has the following consequences:

First, we see that the maximum torque point moves to the left, showing a greater slip. It is interesting to note that the maximum torque value remains unaffected by the rotor resistance even with the shift of the torque curve. The rotor resistance impacts

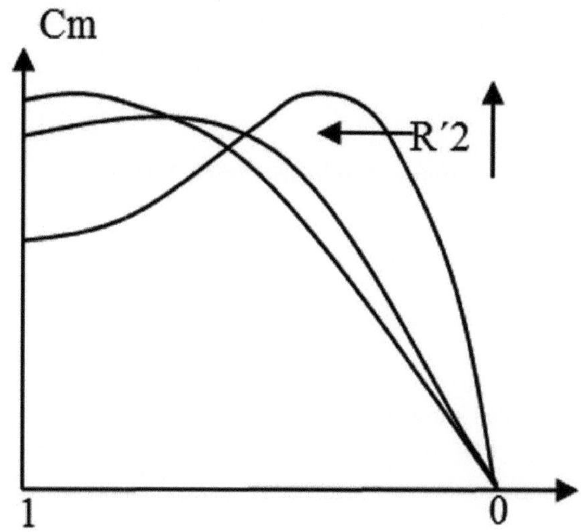

Fig. 9.31 Varying torque with rotor resistance. *Source* own elaboration

the starting torque as it increases. We can plot the operating ranges of asynchronous machines (Fig. 9.32).

Question What is the operation limit?

When the starting torque reaches the maximum torque, the critical point happens. What happen if the rotor resistance increases beyond the limit situation?

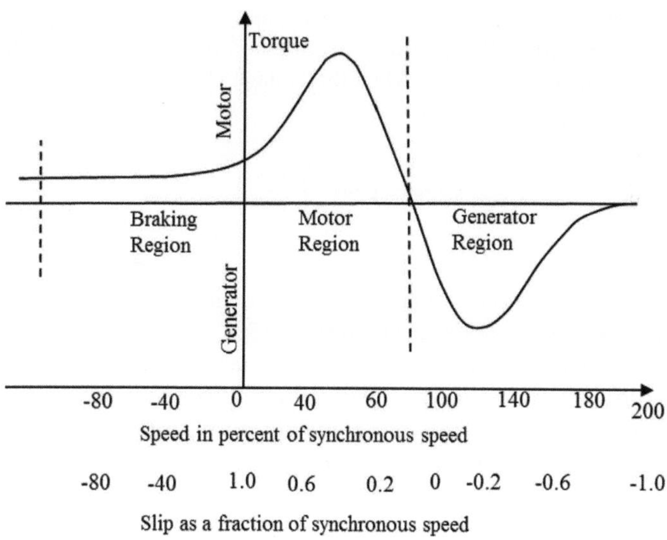

Fig. 9.32 Working regions of asynchronous machine. *Source* own elaboration

This causes a decrease in starting torque of the motor. The maximum torque shift beyond the motor's operating range (s > 1) because of the increase in resistance.

9.11.2 Torque Variation with Voltage

Recalling (Eq. 9.16), we have that

$$\mathrm{Cm} = \frac{\mathrm{P_{mi}}}{2\pi \cdot \mathrm{n_2}} = \frac{\mathrm{P_{mi}}}{2\pi \cdot \mathrm{n_1} \cdot (1 - \mathrm{s})} = \frac{\mathrm{P_{12}}}{2\pi \cdot \mathrm{n_1}} \tag{9.24}$$

We also know that

$$P = \frac{V^2}{R}(W) \tag{9.25}$$

If all the other parameters remain equal, the expression of the torque shows that the value of a specific slip s varies with the square of the voltage.

Keep in mind that:

$$Cm = \frac{P_{mi}}{2\pi \cdot n_2} = \frac{P_{mi}}{2\pi \cdot n_1 \cdot (1 - s)} = \frac{P_{12}}{2\pi \cdot n_1},$$

If the voltage increases while maintaining the frequency value for a certain speed, the motor torque at that speed increases proportionally to the square of the voltage we apply. Thus, if the voltage $\mathrm{V_{1A}}$ produces a specific torque $\mathrm{Cm_A}$ in the asynchronous machine, the torque $\mathrm{Cm_B}$ changes when the voltage changes to $\mathrm{V_{1B}}$. Then, we can write.

$$Cm_B = Cm_A \cdot \left(\frac{V_{1B}}{V_{1A}}\right)^2 \tag{9.26}$$

This relationship holds for any slip.

When the resistant torque is constant, an increase in voltage leads to a reduction in slip, causing a subsequent rise in speed (Fig. 9.33).

On the contrary, increasing the voltage at a constant frequency increases both the flux and the iron losses. As a result, the machine experiences suboptimal current values, if saturation occurs. A decrease in voltage causes the torque the machine generates to decrease, which leads to an insufficient balance against the opposing load. Thus, the motor may stop working.

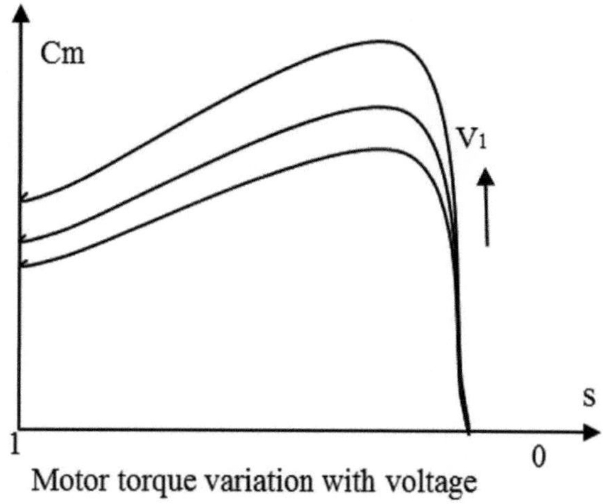

Fig. 9.33 Torque change with voltage. *Source* own elaboration

9.11.3 Torque Variation with Power Frequency

Question How can we avoid over-saturation?

It is common practice to change the frequency while maintaining constant flux in order to mitigate over-saturation risk.

What does it mean to change the frequency while maintaining the flux constant? Then, we know that:

$$\frac{V_1}{f_1} = cte \rightarrow \emptyset_0 = cte \tag{9.27}$$

According to this equation, what happen if the supply voltage and frequency stay proportionate and the flux is constant while the frequency increases? (i e.g., increase from f_{1A} to f_{1B}).

Then we see how torque changes with power frequency in an asynchronous machine (Fig. 9.34). Assuming that the frequency changes (e.g., increases from f_{1A} to f_{1B}), but the flux remains constant, the following happen.

- First, the synchronous speed n_{1A} changes and becomes n_{1B}.
- Secondly, the peak torque remains constant (since the flux remains constant) (Fig. 9.35).

The location of the peak torque reduces. This means a shift to the right. Observe that, for the same curve, the slip equates 1 when the speed is 0, and when the slip is 0, the rotor speed is n_1. The starting torque decreases. We also can state that the torque curve shifts to the right (Fig. 9.35).

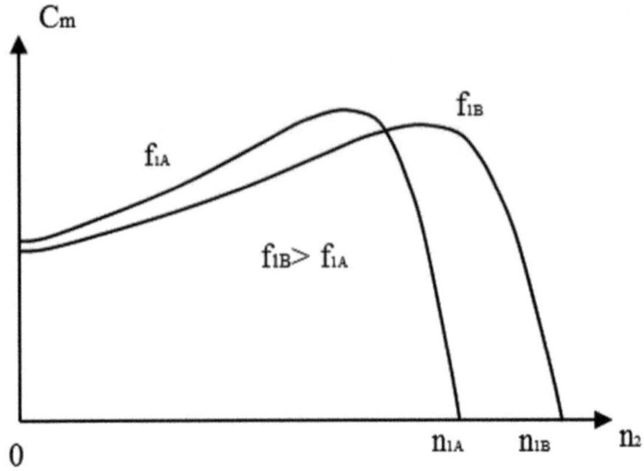

Fig. 9.34 Torque variations with power frequency. *Source* own elaboration

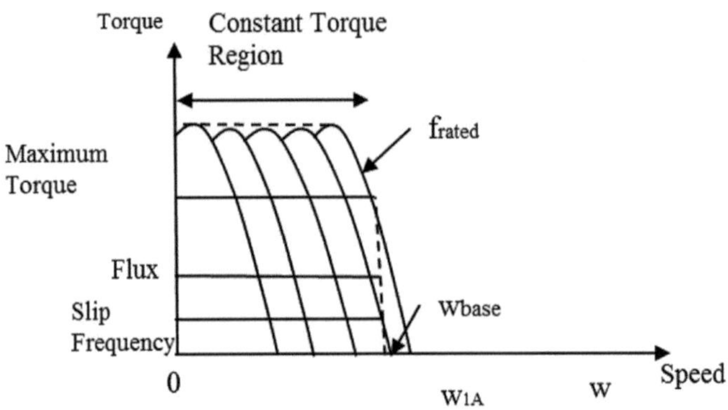

Fig. 9.35 Speed-torque characteristics under constant magnetic flux. *Source* own elaboration

9.12 Asynchronous Motor Operating Zones

Question To reach the operating speed, what had to happen?

In order to achieve the operating speed, the motor torque must be similar to the resisting torque. We plot the operating zones of the asynchronous motor (Fig. 9.36).

The motor has a stable operational region wherein the motor exhibits angular acceleration in response to a load change. The system enters a new state of constant speed as a result of this acceleration. This section looks into how the motor continue running steadily at operating points that correspond to slip values that are less than the maximum slip of torque.

Now, let's assume the resistant torque (C_r) that we see (Fig. 9.36). At points A ($s > s\_C_{max}$) and B ($s < s\_C_{max}$), the motor torque equals the resistant torque. Let's also assume that the motor starts at point B. The motor torque would exceed the resistant

Fig. 9.36 Operating zones in the asynchronous machine. *Source* own elaboration

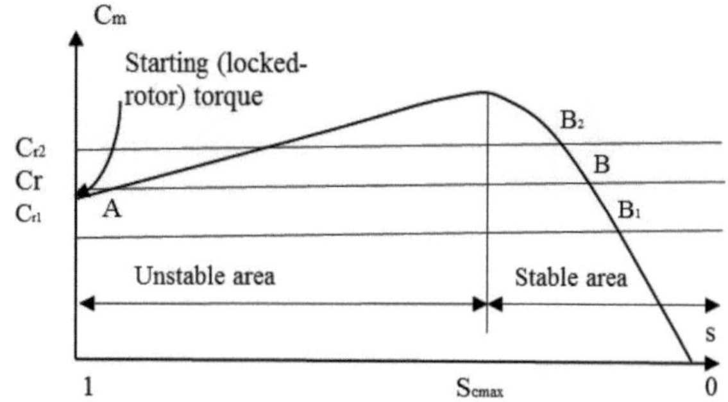

torque if the resistant torque decreases to C_{r1} (as the motor maintains the speed from inertia), resulting in positive acceleration.

The rotor speed would increase (because of less slip), provoking a decrease in the motor torque. The motor would reach a new operating point at B_1, where its torque would match the new resistance torque. If the resistant torque were to increase to C_{r2}, the acceleration would become negative, resulting in a decrease in rotor speed (increased slip), which would lead to a rise in the motor torque. Under these circumstances, B_2 would be the new operating point. We can identify the most important points in the speed-torque curve (Fig. 9.37). Maximum torque point is at rated speed.

We see how the current that the asynchronous machine absorbs affects the operating torque-speed curve (Fig. 9.38).

The figures clearly show that the engine torque passes through a range of values in an uncertain area. Here, if the resistant torque is excessively high and exceeds the torque curve, the machine might have trouble starting or might not start at all. After

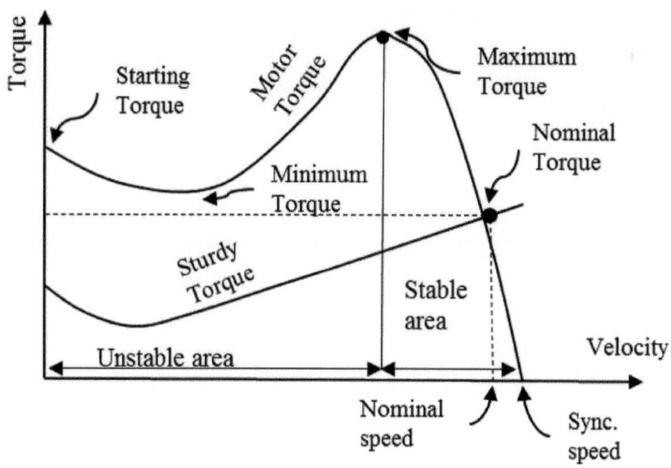

Fig. 9.37 Torque-speed curve. *Source* own elaboration

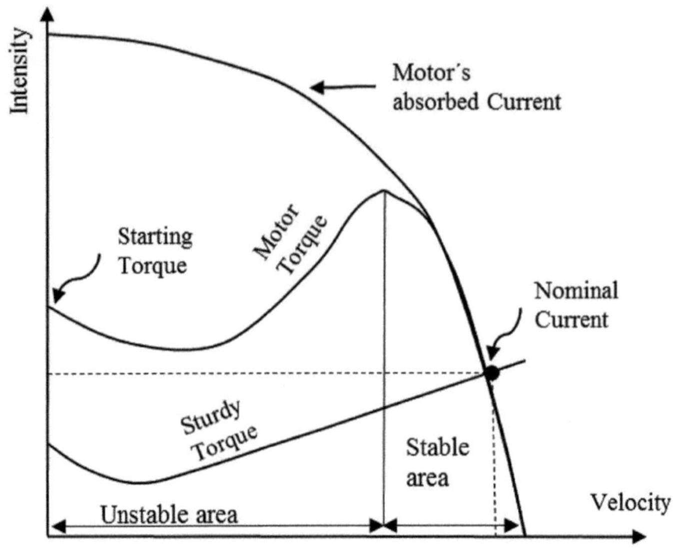

Fig. 9.38 Current-curve over the operating torque-speed curve. *Source* own elaboration

this unstable area is overcome, the engine reaches its designated speed and operates normally.

With a motor, there is an unstable zone where the machine may find problems for starting up, or might be able to start, if the resistant torque exceeds the torque curve. Then, we illustrate the operating areas of an asynchronous machine (Fig. 9.39).

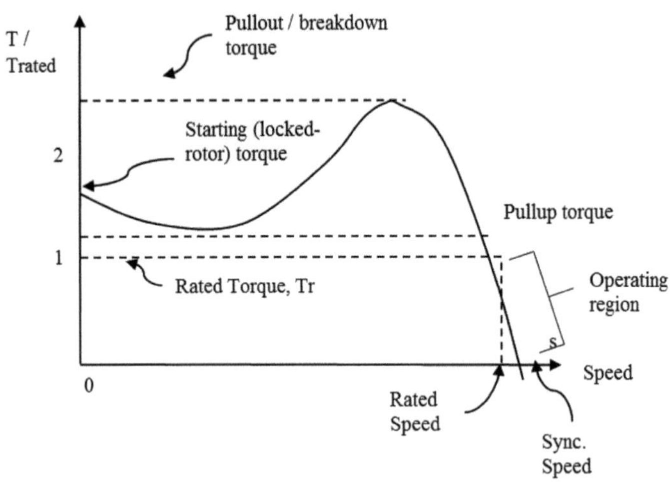

Fig. 9.39 Key points of the torque-speed in an asynchronous machine. *Source* own elaboration

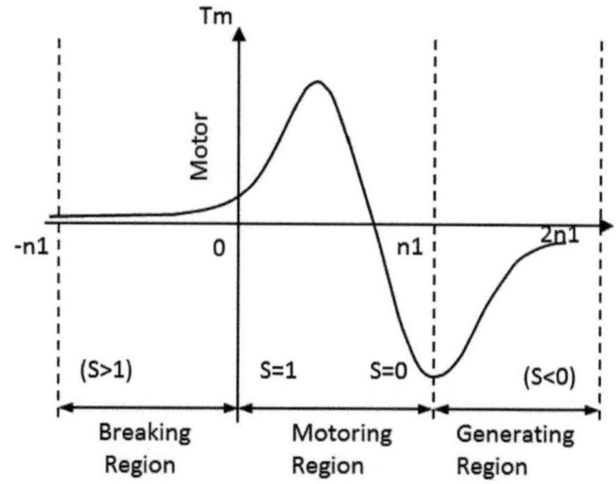

Fig. 9.40 Areas of operation in the torque-speed curve. *Source* own elaboration

Finally, we define the overload capacity of the asynchronous motor as the ratio of the maximum torque to the rated torque. This value typically ranges between 1.6 and 3. Thus far, we evaluated the asynchronous machine acting as a motor. Examining the range of possible operating conditions for this kind of device is a new addition to the study. Thus, the study incorporates the analysis of mechanical power, transmission of stator-to-rotor power, and the motor's torque as a function of slip's variation. Plotting the torque-slip graph for each slip value produces a graph similar to the one shown (Fig. 9.31). It is interesting to plot the areas where the induction machine operates, including the brake, motor, and generator areas (Fig. 9.40).

It demonstrates the different modes of operation the induction machine: motor, brake and generator. We study five different operational scenarios to understand these modes.

At the outset, it is true that: $n_2 = 0$ ($s = 1$), $P_{mi} = 0$; $P_{12} > 0$; $C_m > 0$.

Question What is the value of mechanical power during start-up?

The machine does not supply any mechanical power during this phase since its shaft remains stationary, and the rotor only dissipates power as losses. The following holds true at the no-load point $n_2 \approx n_1$ ($s \approx 0$), with what:

$P_{mi} \approx 0$; $P_{12} \approx 0$; $C_m \approx 0$;

If the slip remains between $0 < s < 1$, the rotor speed varies between zero and the synchronous speed, satisfying:

$P_{mi} > 0$; $P_{12} > 0$; $C_m > 0$.

This means that the rotor receives power P_{12} from the stator and delivers another power, P_{mi}, to the shaft. Under these conditions, the machine operates as an asynchronous motor. In the negative slip zone, the rotor speed exceeds the synchronous speed and both move in the same direction. Note that the rotor speed is higher than the field speed in the air gap.

$$s = \frac{n_1 - n_2}{n_1} \langle 0 \rightarrow n_2 \rangle n_1$$

Consequently, it follows that:

$$Pmi < 0; P12 < 0; Cm < 0$$

This means that the rotor delivers power to the stator (P_{12}) while receives power P_{mi} from the shaft. The machine, therefore, operates as an asynchronous generator. If the slip is greater than one, the rotor speed n2 is in the opposite direction to n1.

As a consequence: $P_{mi} < 0$; $P_{12} > 0$; $C_m > 0$.

This means that the rotor receives power P_{12} from the stator, as well as a certain power, P_{mi}, from the shaft.

Question When the rotor receives power, what would happen?

In this case, the machine operates as a brake, and dissipates all the power that the rotor receives due to Joule losses in its windings.

9.12.1 *Characteristic Curves of the Three-Phase Asynchronous Motor*

Its characteristic graphs indicate the stable operation of an electrical machine. We plot these graphs assuming a constant power supply, voltage and frequency. These conditions are:

$$\frac{n_2}{n_1} = f\left(\frac{P_{\text{useful}}}{P_{\text{rated}}}\right); u = cte; f = cte \tag{9.28}$$

This graph plots the rotor speed as a function of the power that the machine generates. We express speed relative to the synchronous speed, and power to the rated power (Fig. 9.41).

Notice how the speed of the induction motor decreases slightly under load (from 2 to 5% of n_1). Therefore, we can say that the three-phase asynchronous motor possesses a robust speed characteristic.

It is also to highlight other characteristic curves like the consumption characteristic (Fig. 9.42). This graph shows the current drawn from the network as a function of the useful power. The expression of both values is relative to their nominal values.

$$\frac{I_1}{I_{1N}} = f\left(\frac{P_{useful}}{P_{rated}}\right); U = cte; f_1 = cte$$

Note that the current does not drop linearly to zero as P_u gets closer to zero because of the magnetic circuit. When there is no load, the current ranges from 25 to 50% of the rated current. Next graph Fig. 9.43 displays the power factor of the motor concerning the network, depending on the useful power.

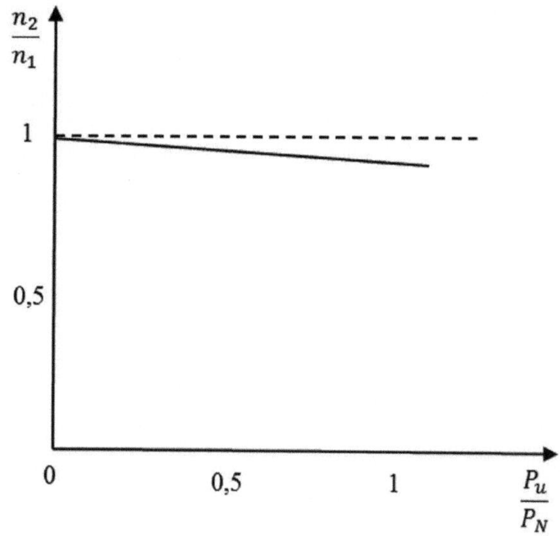

Fig. 9.41 Speed characteristic. *Source* own elaboration

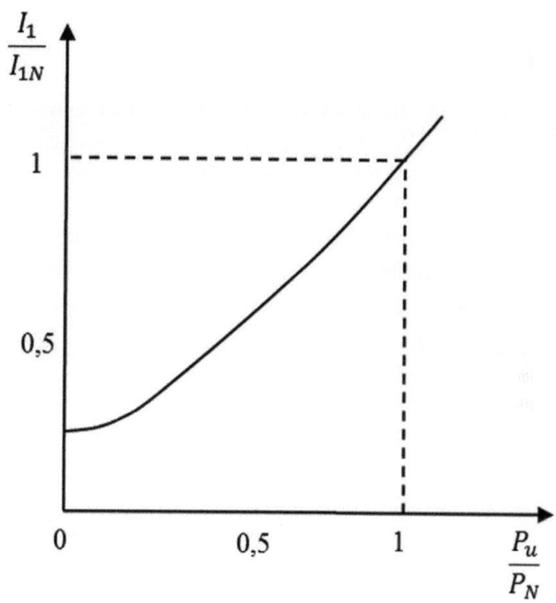

Fig. 9.42 Consumption characteristic. *Source* own elaboration

The power factor is low at small loads (no-load $\cos \varphi \leq 0.2$). However, when the load grows, it rises and drops again when there is an overload.

Question With variable and low loads, what are the measures to take?

Under variable and low loads, a compensating device must improve the power factor. We also highlight the characteristic showing the effectiveness (performance)

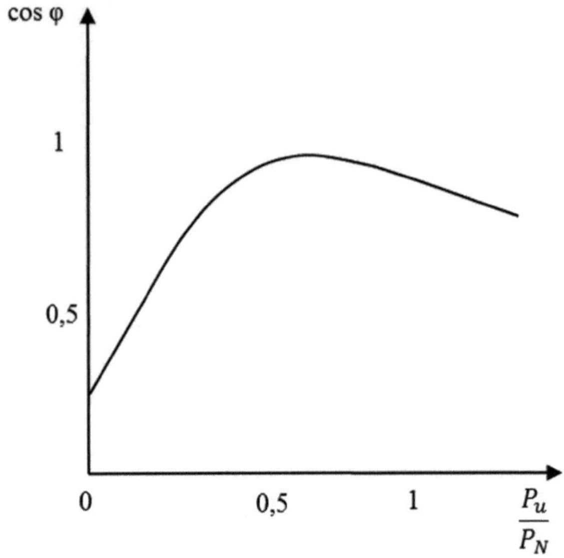

Fig. 9.43 Power factor characteristic of the asynchronous machine. *Source* own elaboration

as a function of the useful power (Fig. 9.44). The efficiency rises swiftly from zero (vacuum) to high levels.

Maximum efficiency occurs when constant losses ($P_R + P_{Fe}$) balance variable losses ($P_{J1} + P_{J2}$) followed by a gradual decrease. The mechanical characteristic evaluates the motor torque as it rotates. It focuses on the relationship between torque and speed (Fig. 9.45).

Since this characteristic curve has already been thoroughly examined, it will not be discussed again.

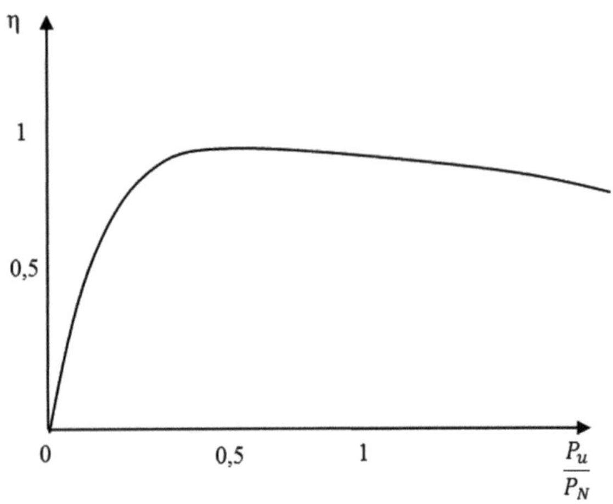

Fig. 9.44 Performance characteristic of an asynchronous machine. *Source* own elaboration

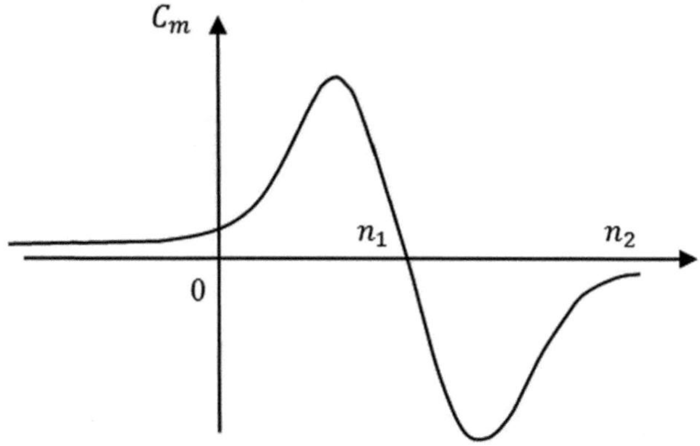

Fig. 9.45 Mechanical characteristic. *Source* own elaboration

9.13 Three-Phase Asynchronous Motor Starting

9.13.1 Introduction. Start-Up Methods Needed

We see that plotting the torque-speed graph of a three-phase induction motor is essential to understand what the starting torque is (Fig. 9.46).

In the torque-speed graph of a three-phase induction motor we see the starting torque. The starting torque of the motor must be greater than the resistant torque (Cr, C_{r1}, C_{r2}) that the mechanical load attached to the shaft imposes. Note that, during start-up, the rotor speed is zero ($n_2 = 0$), the slip is 1, and the starting resistance of the rotor winding is zero. This point corresponds to a short-circuit scenario (Fig. 9.47). It is a transient state because the current decreases as the shaft speed rises.

Question What is the current that flow through the motor during startup?

During the transient, a short circuit occurs, causing high currents (8–10 times the rated current) to flow.

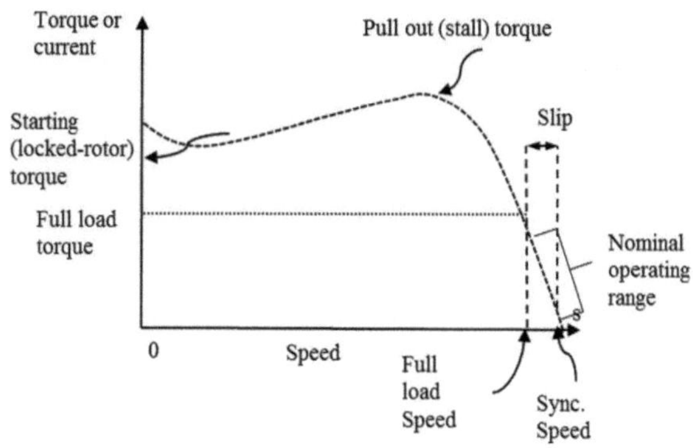

Fig. 9.46 Mechanic characteristic. *Source* own elaboration

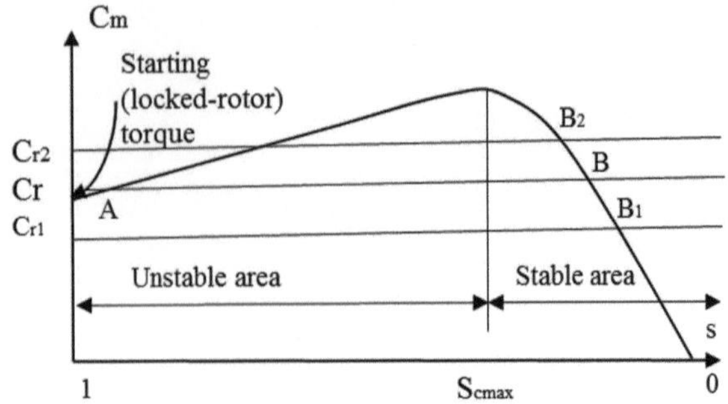

Fig. 9.47 Starting torque in the mechanical characteristic. *Source* own elaboration

Question Does the high current during startup have any consequence?

High current results not only in significant Joule losses in the motor but also can lead to substantial voltage drops in the supply lines, which can affect other loads that connect to the same line. It is common that systems limit the current. It is of interest to show the values for IEC 60,634 countries (Table 9.1).

In order to satisfy these requirements, it is common to use methods that reduce the current drawn by the motor. It is crucial to minimize the current that the motor consumes during start-up in order to comply with the regulations in different countries (Table 9.1). The current legislation stipulates the highest permissible levels for the motor outset. Hence, it is important to lower the current that the motor consumes during start-up to comply with the regulations.

Question What methods do we employ to lower current during start-up?

The majority of current reduction starting techniques include lowering the motor's supply voltage while it is starting. This is because when a motor is starting-up, the supply voltage increases to reduce the inrush.

Table 9.1 Maximum values of start current

Power of motor	Istart/Inominal
From 0.75 kW to 1.5 kW	4.5
From 1.5 kW to 5 kW	3
From 5 to 15 kW	2
More than 15 kW	1.5

Source own elaboration

9.14 Summary

This chapter provides a comprehensive overview of asynchronous machines, which are a cornerstone of electromechanical energy conversion. It begins by classifying these machines based on their number of phases, number of poles, and rotor type—primarily distinguishing between the squirrel-cage rotor and the wound rotor.

The core of the chapter explains their operating principle, centered on the concept of the rotating magnetic field (supported by foundational experiments like Arago's and the Ferraris theorem). It details the various operating modes (motor, generator, brake) but focuses extensively on their most common use as a motor.

Key performance concepts are explored, including the comparison of an asynchronous motor to a transformer, the analysis of power distribution, efficiency, and torque. The chapter concludes by examining the motor's characteristic curves, operating zones, and the different methods required for starting a three-phase asynchronous motor.

References

1. Wikipedia. Schnittmodell durch einen heute üblichen Drehstrommotor, ein sogenannter Asynchronmotor. Im Gegensatz zum Synchronmotor ist die Rotordrehzahl um den Faktor des Schlupfs gegenüber dem vom Drehstrom erzeugten Drehfeld reduziert. Anwendungen liegen bei Antrieben wo das drehzahlabhängige Drehmoment nicht besonders stört, wie zum Beispiel bei Pumpen oder Ventilatoren (2011). https://commons.wikimedia.org/wiki/File:TMW_50906_Schnittmodell_einer_Drehstrommaschine_(Asynchronmaschine).jpg. Accessed 7 Dec 2023
2. P. Loos, Coupe du rotor d'une machine asynchrone (2005). https://commons.wikimedia.org/wiki/File:Coupe_rotor_machine_asynchrone.jpg. Accessed 8 Jan 2024
3. Biezl, Deutsch: zwei poliger de:Schleifringläufermotor mit drei Strängen. https://commons.wikimedia.org/wiki/File:Funktionsprinzip-schleifringl%C3%A4ufer.svg. Accessed 7 Dec 2023
4. «Файл:SL742199.JPG—Уикипедия». https://commons.wikimedia.org/wiki/File:SL742199.JPG. Accessed 7 Dec 2023
5. R. B. Slovakia Pezinok, Induction motor squirel cage illustration (2006). https://commons.wikimedia.org/wiki/File:AM_Klietka.png. Accessed 7 Dec 2023
6. Zureks, Stator of an electric motor with windings (2006). https://commons.wikimedia.org/wiki/File:Uzwojenie.jpg. Accessed 7 Dec 2023
7. R. Electrician, Servicing the asynchronous motor of the Yauza-209-1 tape recorder. Rotor with two integral fans (2021). https://commons.wikimedia.org/wiki/File:Yauza_209-1_(motor)_-_05_-_rotor_with_fans.jpg. Accessed 7 Dec 2023
8. R. Electrician, Servicing the asynchronous motor of the Yauza-209-1 tape recorder. Stripdown: stator, rotor,top cap and two sets of bearings with their grease seals. Here, they have already been washed and degreased (2021). https://commons.wikimedia.org/wiki/File:Yauza_209-1_(motor)_-_04_-_motor_disassembled.jpg?uselang=es. Accessed 7 Dec 2023
9. Wikipedia. Rechte Hand Regel (2008). https://commons.wikimedia.org/wiki/File:RHR.svg?uselang=es. Accessed 7 Dec 2023
10. MdeVicente: Rotating magnetic fieldEspañol: Campo magnético rotatorio, teorema de Ferraris (2014). https://commons.wikimedia.org/wiki/File:Campo_magnetico_rotatorio.jpg. Accessed 7 Dec 2023

11. Wikipedia. Stator feuilleté.jpg. https://commons.wikimedia.org/wiki/File:Stator_feuillet%C3%A9.jpg. Accessed 7 Dec 2023
12. Wikipedia. Rotating-3-phase-magnetic-field.svg. https://commons.wikimedia.org/wiki/File:Rotating-3-phase-magnetic-field.svg. Accessed 2 Jan 2024

Chapter 10
The Synchronous Machine

Abstract The primary purpose of the synchronous machine is to generate electricity. The synchronous generator contributes to power system stability because of its built-in ability to operate at a steady frequency. We begin the study of these machines reviewing the components of the synchronous machine. According to the synchronous speed, various configurations emerge from this knowledge. This chapter also shows the operation of the asynchronous machine by analyzing the characteristics curves of the synchronous machine.

Keywords Synchronous machine · Power generation · Synchronous machine parts · Synchronous generator under load · Reactive power · Power factor correction

10.1 Introduction

The main characteristic of synchronous machines is that they rotate at a constant speed. Synchronous speed (n) is the name we give to this constant speed. The machine's number of poles (2p) and network frequency (f) determines this speed. We can express it as:

$$\text{Synchronous speed} = \text{n(rpm)}\frac{60 \cdot f}{p} \tag{10.1}$$

Depending on the direction of energy conversion -from mechanical to electrical or electrical to mechanical- these devices can be motors or generators. The most important application is the electrical generator. Large power plants with synchronous generators are responsible for maintaining the electrical grid. In synchronous machines, alternating current powers the stator winding while direct current powers the rotor winding.

D. Borge-Diez et al., *Fundamentals of Electrical Power Engineering*, Lecture Notes in Electrical Engineering 1540, https://doi.org/10.1007/978-3-032-18904-2_10

10.2 Constituent Elements of the Synchronous Machine

The following is a brief description of the most relevant components of this machine.

10.2.1 Rotor

The rotor is part of the magnetic circuit of the machine. It can be cylindrical or salient pole-shaped rotor Fig. 10.1a and b.

Diagram of a cylindrical electric motor with labeled components. The outer circle represents the stator with three labeled connections: U, V, and W. The inner circle shows the rotor with a magnetic field indicated by "S" for the south pole and "" for the north pole. A small circuit symbol is depicted at the center of the rotor. The motor is supported by two stands.

Another configuration is a rotor with protruding poles (Fig. 10.2). Because of both friction and centrifugal forces that suffer the poles, rotational speeds in this type of rotor must be lower than 1,000 rpm.

The windings attach to the poles. This makes it necessary, for a given frequency, to increase the number of poles and, to use large diameter rotors. The pole attachment must be strong enough to balance the effects of centrifugal and magnetic forces. Cylindrical rotors can rotate at high speeds (above 1,000 rpm). In this case, they

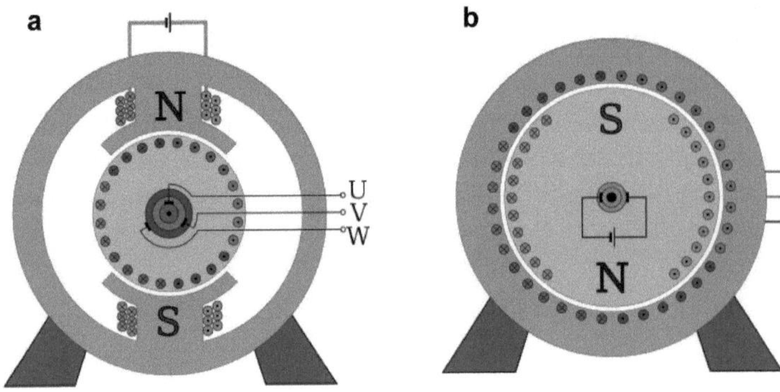

Fig. 10.1 **a** and **b** Salient pole and cylindrical rotor of the synchronous machine [1]

Fig. 10.2 Salient-pole rotor of a synchronous machine. *Source* own elaboration

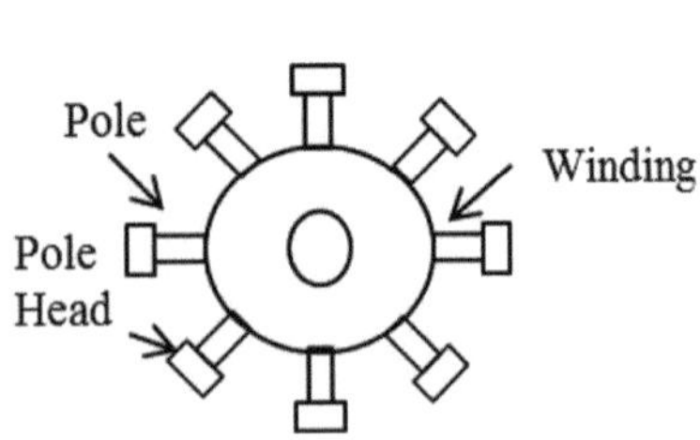

Fig. 10.3 Synchronous machine. Original from Alexander Seetenky. *Source* from wikicommons under CC3.0 to use license [2]

consist of a magnetic material cylinder with grooves around the outer perimeter to house the electrical winding. In power generation, it is common to find cylindrical rotor power generator (Fig. 10.3).

Compared to salient pole rotors, this design results in rotors that are longer and have a smaller diameter than salient pole rotors.

10.2.2 Stator

It is made of a set of stacked and separately insulated plates. The internal perimeter of the plates features grooves (Fig. 10.4) where the conductors locate. Additionally, the machine cools through a few holes or ventilation channels.

In order to reduce iron losses in the machine's magnetic circuit and enhance performance, these sheets are stacked and insulated from one another (Fig. 10.5).

10.2.3 Stator Winding

This is the electrical circuit in the stator of the synchronous machine. It comprises turns (Fig. 10.6) which, depending on the number of phases and poles of the generator, locate along the stator magnetic package slots as previously showed.

Synchronous generators (Fig. 10.7) often have two pairs of poles and an inductor with permanent magnets.

The figure below Fig. 10.8 shows a single voltage cycle of a three-phase system with labels along the time axis from 0 to 360° (2 π radians). The graph shows how the voltage or current varies over time. The cycle repeats 50 or 60 times per second, depending on the power system frequency.

When the generator works under load the circuits with coil, or pure inductance, alternating voltages appear in these loops.

Fig. 10.4 Sheets of stator in synchronous machine. *Source* from wikicommons under free to use license [3]

Fig. 10.5 Stator of synchronous machine. *Source* from wikicommons under free to use license [4]

10.2.4 Rotor Windings

It is the rotor's electrical circuit. The function of this winding is to generate the inductive poles of the machine using direct current. Manufacturing this winding varies depending on whether it is a cylindrical rotor or a salient pole rotor. As previously mentioned, we have different arrangements of salient-pole machines (Fig. 10.9). In

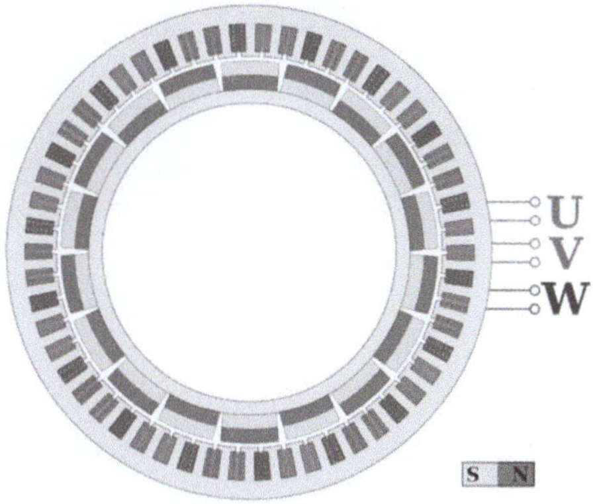

Fig. 10.6 Multiphase synchronous machine. *Source* from wikicommons under free to use license [5]

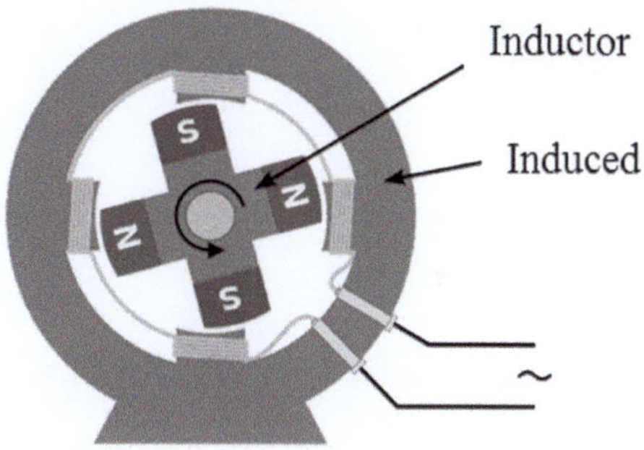

Fig. 10.7 Synchronous generator with two pairs of poles and an inductor with permanent magnets. *Source* from wikicommons under free to use license [6]

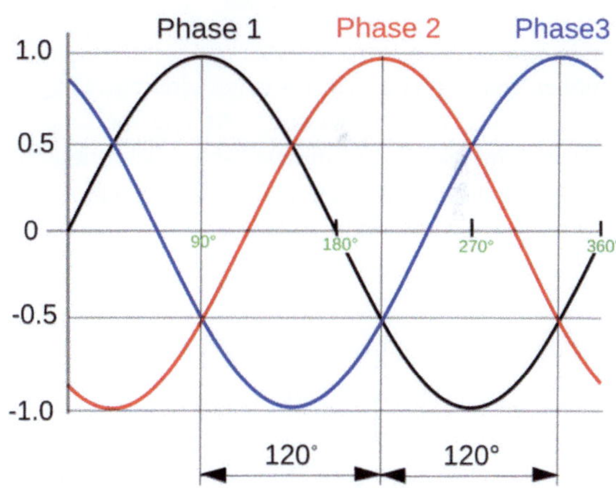

Fig. 10.8 Three-phase sine wave. Original. *Source* from wikicommons under CC3.0 to use license [7]

cylindrical rotor machines, we house the windings in slots. The damper winding is in the outermost part of the pole expansions of the salient pole synchronous machines. It comprises copper bars that house in grooves as close as possible to the air gap. These bars are riveted at both ends to segments or rings that close the winding. The

Fig. 10.9 Salient poles with damper winding. *Source* from wikicommons under free to use license [8]

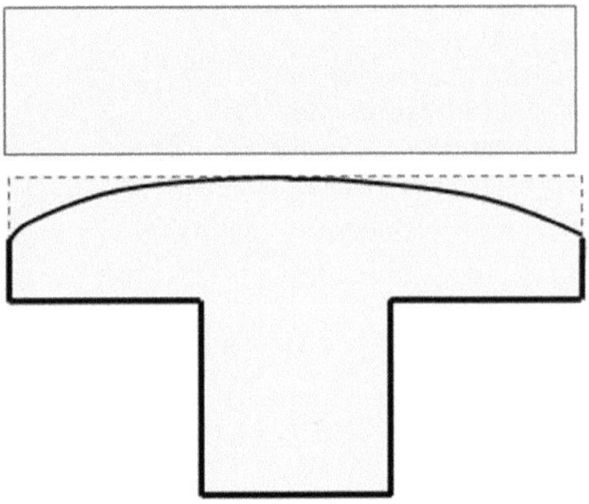

Fig. 10.10 Salient-pole detail. *Source* own elaboration

snubber winding is a squirrel cage winding. The solid mass of a cylindrical rotor also acts as a damping winding. Among the functions it performs is to highlight: reducing emf harmonics, maintaining synchronous speed, damping variations in the magnetic field; and starting the synchronous motors.

The variable air gap of salient pole machines induces variations in the magnetic field (Fig. 10.10).

10.2.5 Excitation System

DC supplies the electricity of the rotor winding. Currently, it is common to use two different technologies.

- The static excitation system, and the
- Brushless excitation system.

Rectifier equipment generates (Fig. 10.11), the direct current required to feed the rotor winding based on power electronics. After that, the electric winding receives

Fig. 10.11 Rotor windings fed with direct current. *Source* from wikicommons under free to use license [9]

this electricity via a system of brushes rubbing against rotating rings that work with the shaft and rotor winding.

In contrast, the brushless excitation system generates direct current via an inverse generator (comprising an inductor in the stator and armature in the rotor). This generator produces alternating current, and the diodes rectify the current. Then, it rotates in conjunction with the rotor shaft and its electrical winding.

10.2.6 Cooling System

The machine contains multiple heat sources, including eddy current losses and the Joule effect. Therefore, we need a heat dissipation system to keep the machine's parts from aging too. We can illustrate a schematic of the system which includes the cooling system (Fig. 10.12).

The most common internal cooling medium of the machine is air, which flows through the ventilation ducts, cooling both the rotor and the stator.

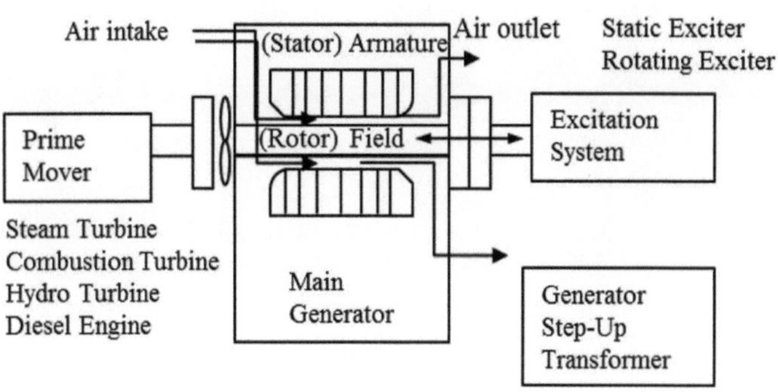

Fig. 10.12 Synchronous machine system. *Source* own elaboration

10.2.7 Outer Casing

Inside the housing we find the rotor and the stator. The exciter (Front left) and the generator at the back are covered with the outer casing shown in the Fig. 10.13.

Question Why are both the stator and the rotor inside this housing?

The purpose of the casing is to ensure the machine's stability, fixation, and protection of its constituent parts from the outside world.

10.3 Elementary Synchronous Generator

The basic synchronous generator includes a stationary inductor in the stator and a movable armature in the rotor. The most elementary arrangement consists of two-pole elementary generator (Fig. 10.14).

Examining a basic generator, we find that it consists of a cylindrical rotor with a diameter D and an axial length L, matching the depth of the poles, and a stator or inductor with two poles, north and south. It is of vital importance to understand the distance in windings (Fig. 10.15).

In the rotor (Fig. 10.16), we have a loop comprising two conductors, AA' and BB' that connects to a conductor $A'B'$. This conductor is not within the field of influence of the magnetic field.

When the rotor rotates at n rps, the linear velocity of loops AA' and BB' is:

$$v = \omega \cdot radio = 2\pi \cdot n \cdot \frac{D}{2} = \pi \cdot n \cdot D \tag{10.2}$$

Fig. 10.13 Outer casing. *Source* from wikicommons under free to use license [10]

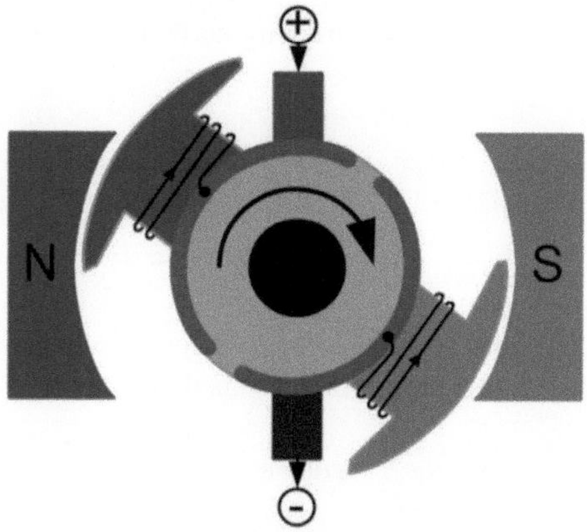

Fig. 10.14 Stator core and two-pole rotor. Original from haade. *Source* original from wikicommons under CC3.0 use license [11]

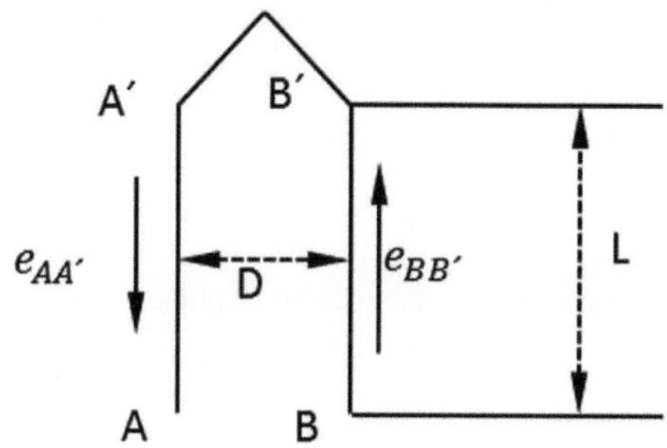

Fig. 10.15 Elementary synchronous generator stator windings. *Source* own elaboration

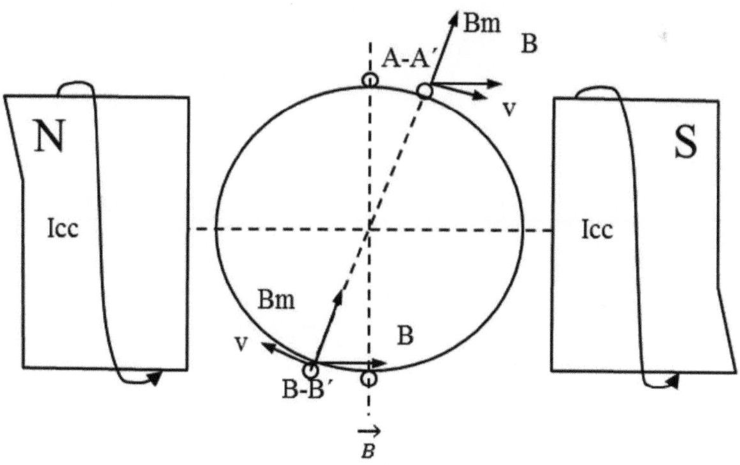

Fig. 10.16 Rotor wiring scheme. *Source* own elaboration

According to Faraday's law, when moving coils AA' and BB' within a magnetic field, induced voltages are:

$$e_{AA'} = v \cdot B \cdot L \cdot sen\theta \tag{10.3}$$

$$e_{BB'} = v \cdot B \cdot L \cdot sen(\pi - \theta) \tag{10.4}$$

Keep in mind that both voltages have the same magnitude and direction (Fig. 10.17). The induced emf within the loop is then:

$$e_{AB} = 2 \cdot e_{AA'} = 2 \cdot v \cdot B \cdot L \cdot sen\theta = 2\pi \cdot n \cdot (D \cdot B \cdot L) \cdot sen\theta \tag{10.5}$$

The flux that the loop embraces is at its maximum value when it is vertical and perpendicular to the lines of the magnetic field and becomes:

$$\Phi_0 = B \cdot D \cdot L \tag{10.6}$$

D-L is the part that is inside the loop at the position of maximum flow.

Therefore, it is true that:

$$e_{AB} = 2\pi \cdot n \cdot \Phi_0 \cdot sen\theta \tag{10.7}$$

The total emf in the N loops is:

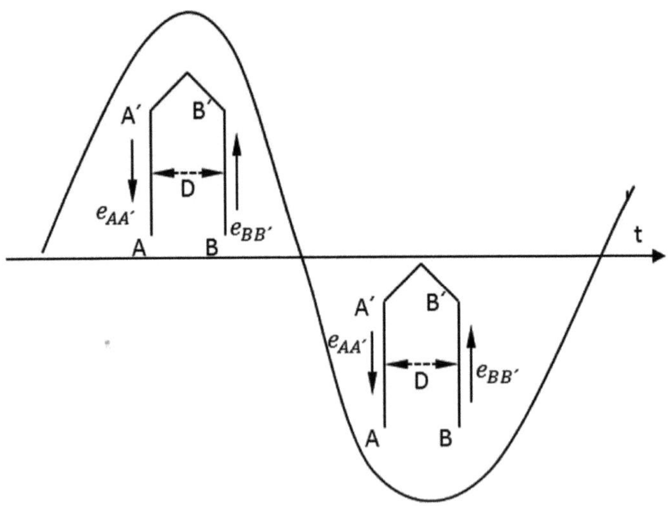

Fig. 10.17 Basic synchronous generator stator. *Source* own elaboration

$$e = N \cdot 2\pi \cdot n \cdot \Phi_0 \cdot sen\theta \tag{10.8}$$

Recalling the definition of angular displacement θ, we have:

$$\theta = w \cdot t = 2 \cdot \pi \cdot n \cdot t$$

Therefore, in the case of N turns, the emf is a sinusoidal function of magnitude. Substituting this equation into (Eq. 10.9) we have that:

$$e = N \cdot 2\pi \cdot n \cdot \Phi_0 \cdot sen(2\pi \cdot n \cdot t) \tag{10.9}$$

Upon a comprehensive analysis of a rotor's complete rotation, it becomes evident that loops AA' and BB' change polarity, depending on whether they are under the influence of the North or South Pole.

Each turn guarantees a complete emf cycle. It is important to note that the coils have a polarity during the first half-cycle, and the opposite in the second half-cycle.

The value of this emf is then:

$$E = \frac{2 \cdot \pi \cdot N \cdot f \cdot \Phi_0}{\sqrt{2}} = 4.44 \cdot N \cdot \Phi_0 \cdot f \tag{10.10}$$

10.4 Real Synchronous Generator

An authentic synchronous generator has the inductor in the rotor and the armature in the stator. In a generator with a set of poles, the direct current that flows through the rotor winding establishes the poles, designating the north and south poles based on the current flow (Fig. 10.18). The rotor-air gap-stator path encloses the magnetic field force lines.

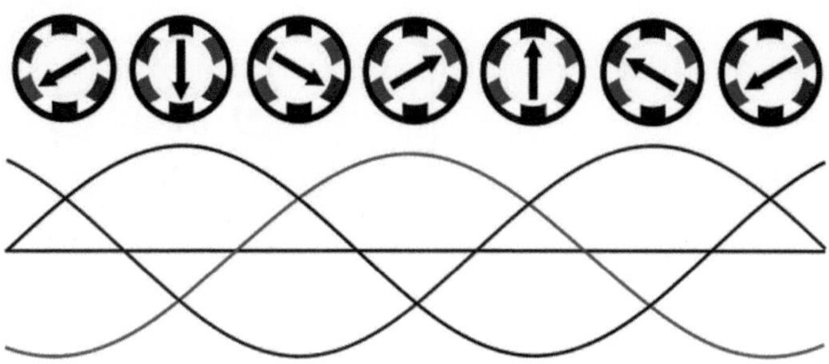

Fig. 10.18 Synchronous generator induced waves. *Source* from wikicommons under free to use license [12]

Question What do we get if we use an external drive mechanism to rotate the poles?

As we know, rotating the poles by means of an external drive, like a turbine, it appears a rotating magnetic field.

Question What does represent a rotating magnetic field in a real synchronous generator?

A rotating magnetic field moves with respect the stator coils. The application of Faraday's law results in the generation of an emf in the stator coils. These conductors do not move. Faraday's law requires taking the conductor's velocity regarding the field ($A-A'$) into account in order to determine the polarities of the induced electromotive forces in the stator conductors. However, in the machine, it is the field that moves relative to the conductor (v).We see that rotor's rotation causes a polarity change of the induced electromotive forces in the windings (Fig. 10.19). Specifically, the forces enter under the North Pole and exit under the South Pole (in the rotation direction).

According to Faraday's law, we have to plot relationship between coil speed and electromagnetic forces (Fig. 10.20).

Diagram illustrating electromagnetic induction. A straight arrow points diagonally upward, labeled "Coil velocity relative to the field." A curved arrow labeled "" and "" indicates the direction of rotation, with the text "polarity of the induced emf" nearby. The sketch represents the relationship between coil movement and induced electromotive force.

The rotation of the stator will induce a total emf in them.

$$E = 4,44 \cdot N \cdot \Phi_0 \cdot f \tag{10.11}$$

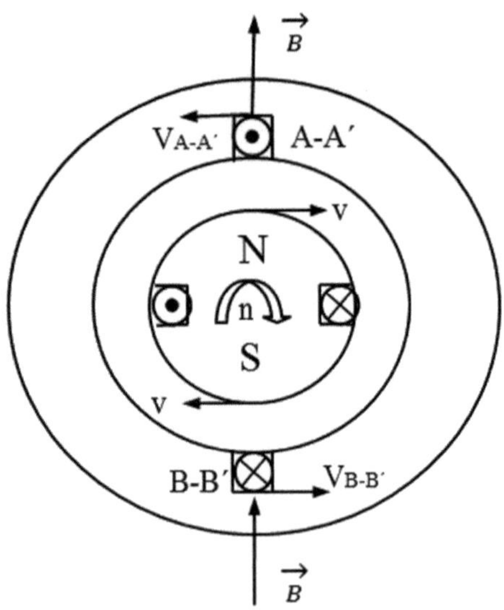

Fig. 10.19 Polarity of induced electromagnetic forces. *Source* own elaboration

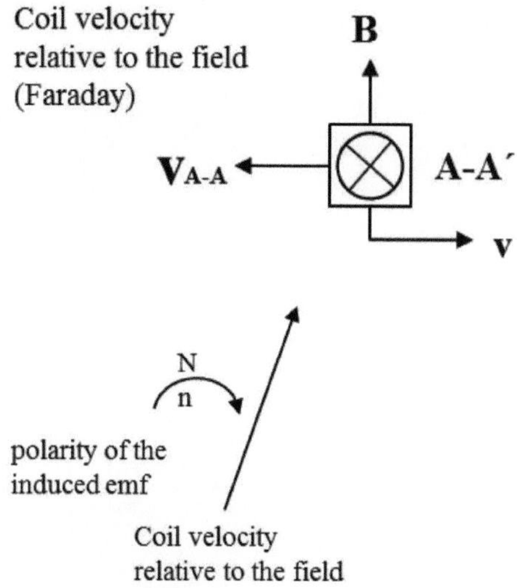

Fig. 10.20 Relationship between coil speed and electromagnetic forces. *Source* own elaboration

As with transformer windings, and asynchronous machines these turns have some resistance and reactance.

10.4.1 Types of Real Synchronous Machines

Remember that there are two constructive rotor types: the salient-pole rotor, also known as variable air gap rotor (or variable reluctance), and the cylindrical rotor, also known as constant air gap rotor (Fig. 10.21). In both cases, we assume sinusoidal magnitudes of magnetic induction, flux and reluctance.

Irrespective of the kind of the synchronous machine, the rotating magnetic flux interferes with the coils, and it appears a sinusoidal waveform (Fig. 10.22).

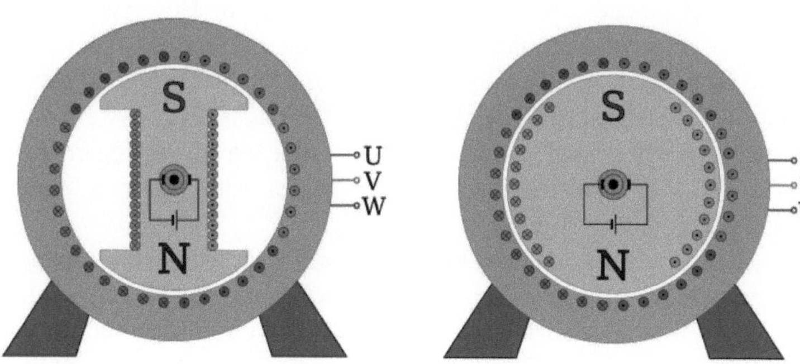

Fig. 10.21 Rotor types in synchronous generators. *Source* from wikicommons under free to use license [13, 14]

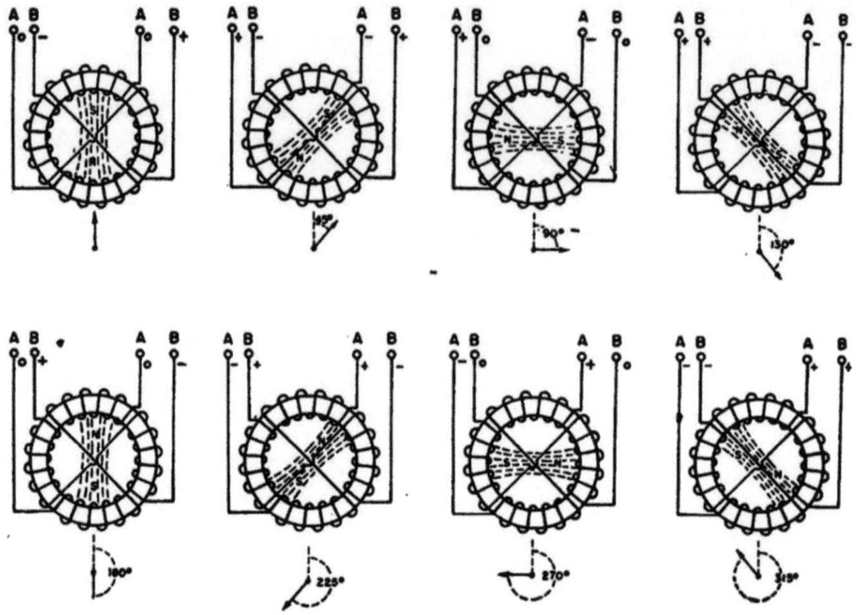

Fig. 10.22 Rotating magnetic flux. *Source* from wikicommons under free to use license [15]

Therefore, the machine generates a sinusoidal magnetic induction wave that travels along the machine as a result of the rotor's rotation. Note that at any point on the stator (e.g., at conductor AA'), the value of the induction is different because the shaft changes as the rotor rotates. We see that there is a sinusoidal induction wave throughout the machine. The rotor creates this wave. As the rotor moves, the induction wave moves with it, so that the induction at each point on the stator varies with time. We see the rotating magnetic flux that appears (Fig. 10.23).

Therefore, depending on the number of poles, different speeds of the rotating magnetic flux speed appear. In this sense, with gas turbines, their performance requires very high speeds, which can reach 3.000 rpm. However, in hydraulic turbines, their performance requires a low rpm, being impossible to get 3.000 rpm. It is also helpful to know how revolutions per minute for 50 Hz and 60 Hz correlate with the number of pole pairs Table 10.1.

The synchronous machine produces a sinusoidal waveform regardless of the number of poles. It is interesting to plot the waveform that a 4-pole machine, or a 2-pole machine generate (Fig. 10.24).

Similarly, a two-pole synchronous machine (Fig. 10.25), generates a voltage wave.

In order to achieve a frequency of 50 Hz in a generator with rotational speeds below 3000 rpm, it is necessary to equip the rotor with a greater number of poles. For example, in the case where there are two pole pairs, the induced electromagnetic forces cycle twice for every complete rotation of the rotor. This happens as a result of the fact that during each complete rotor rotation, two north and two south poles

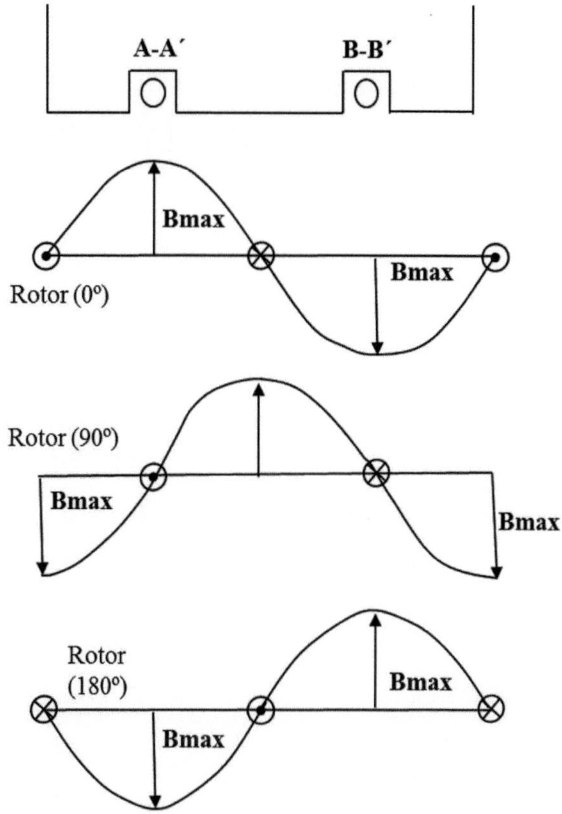

Fig. 10.23 Induced waveform. *Source* own elaboration

Table 10.1 Correlation between the number of pole pairs and revolutions per minute

Pole-pairs	For 60 Hz rev/min (N)	For 50 Hz rev/min (N)
1	3600	3000
2	1800	1500
3	1200	1000
4	900	750

Source own elaboration

alternately pass in front of each stator conductor. If the generator has p pairs of poles, each 360° rotation of the rotor results in p cycles that the electromotive forces induce.

As seen above, it is satisfied that:

$$\theta \text{ electric} = p\,\theta \text{ geometric} \tag{10.12}$$

The angular separation between the successive north and south poles amply shows this expression.

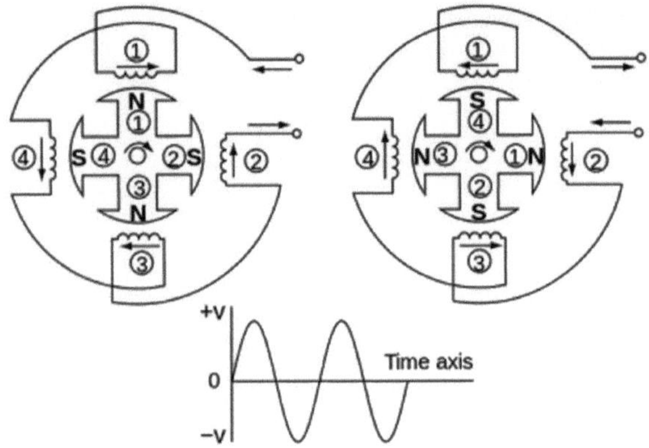

Fig. 10.24 4-pole synchronous machine. *Source* from wikicommons under free to use license [16]

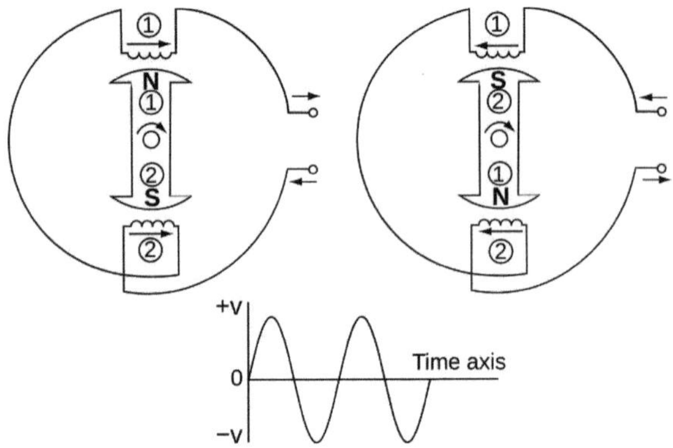

Fig. 10.25 2-pole synchronous machine. *Source* from wikicommons under free to use license [17]

$$p \cdot n = 60 \cdot f \tag{10.13}$$

Between the North and South Pole (consecutive north and south) there must be a magnetic distance of 180º (magnetic degrees are equivalent to electrical degrees) which is achieved with 180º geometric with one pair of poles, or with 90º geometric with two pairs of poles.

10.4.2 Multiphase System

So far, we have only considered single-phase generators. With three-phase generators, it is necessary to get a balanced three-phase voltage. This arrangement gives us a 120º electrical degrees voltage system (Fig. 10.26).

To implement the multiphase system in the synchronous machine, the poles have the same behavior, inducing energy in all of them. The rotor with the poles induces the flux in the multiphase system (Fig. 10.27).

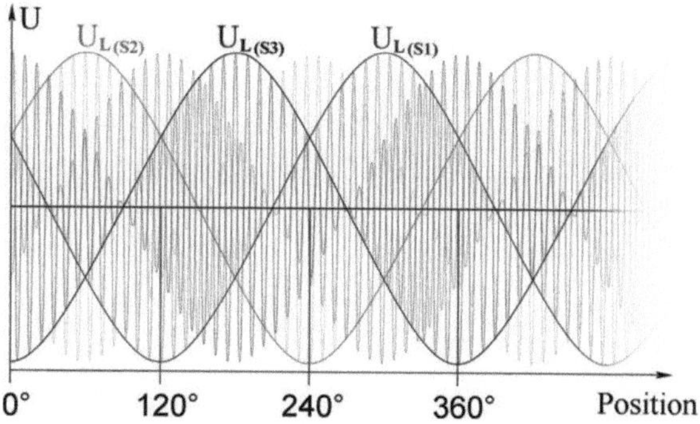

Fig. 10.26 3-Phase voltage balanced system. *Source* from wikicommons under CC3.0 use license [18]

Fig. 10.27 Poles of synchronous machine. Stator of an electric machine with windings [19]

Therefore, the loops in the machine must locate on the stator for the rotating field induces electromotive forces with electrical phase shifts of 120º. Therefore, if at a given moment someone places a north pole on a conductor of phase 1, after 120º electrical (the geometrical correspondence must be seen), they must place it on a conductor of phase 2.

10.5 No-load Synchronous Generator

It is useful to distinguish between no-load and load scenarios when we analyze the voltage that the synchronous generator generates. A synchronous generator is in a no-load condition when its armature winding is disconnected. We see the functional diagram of a q-phase and no-loaded generator (Fig. 10.28).

Under these conditions, the magnetomotive force (Fmme) appears only because of the direct current (I_e) circulating through the inductor. This excitation generates

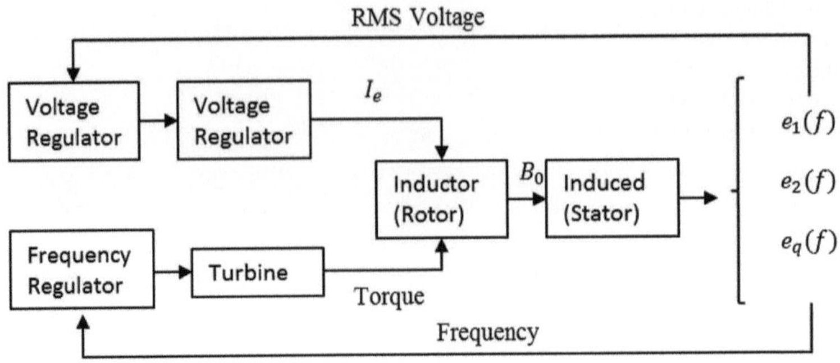

Fig. 10.28 Functional diagram of a q-phase no-load alternator. *Source* own elaboration

flux circulation in the magnetic circuit of the synchronous machine and produces voltage e (t) in the stator's armature winding. Assuming the inductor generates a sinusoidal induction wave, the voltage in the armature winding is also sinusoidal. As previously mentioned, the frequency of the induced voltage is:

$$f = \frac{p \cdot n}{60} \tag{10.14}$$

The following expression provides the rms value of the induced no-load voltage e (t):

$$E_0 = 4,44 \cdot N \cdot \Phi_0 \cdot f \tag{10.15}$$

It is important to note that, due to the absence of currents in the armature, the voltage is the same that the phase voltage (V) at the generator terminals.

10.6 Synchronous Generator Under Load

The synchronous generator operates under load when the armature circuit closes over an external electrical load. This provokes that voltage system in the armature power the external electrical loads. The current flows through the armature coils when the generator is under load, and the voltage at the generator terminals (V) differs from that in the no-load situation (E_0) with the same excitation current. The voltage regulation coefficient (ε), which we define now on measures the voltage fluctuation.

$$\varepsilon = \frac{E_0 - V}{V} \cdot 100 \tag{10.16}$$

When a generator operates under load conditions; the force of magnetomotive torque opposes the rotational movement of the shaft (Fig. 10.29). The generator operates through the use of water turbines, gas turbines, and other means of driving the shaft, and supplies current once connects to a load.

The current flowing (which can be single-phase, three-phase, or multi-phase) through the stator provokes the magnetomotive force: it is the armature reaction. Under these conditions the resulting flux in the machine because of the magnetomotive force (F_{mme}) (Fig. 10.30) and the armature reaction magnetomotive force (F_{mmi}). The flow of a balanced three-phase system of currents through the armature (stator) feeding balanced three-phase loads (which is how generators work) produces a rotating speed field. We define it (Eq. 10.17) as:

$$n_{\text{induced reaction}} = \frac{60 \cdot f}{p_{\text{stator}}} \tag{10.17}$$

Taking into account the frequency of the currents passing through the armature and the fact that the generators have the same number of poles in both the stator and the rotor, we deduce Eq. 10.18.

$$n_{induced\ reaction} = \frac{60 \cdot f}{p_{stator}} = \frac{60 \cdot \frac{p \cdot n}{60}}{p_{stator}} = n \tag{10.18}$$

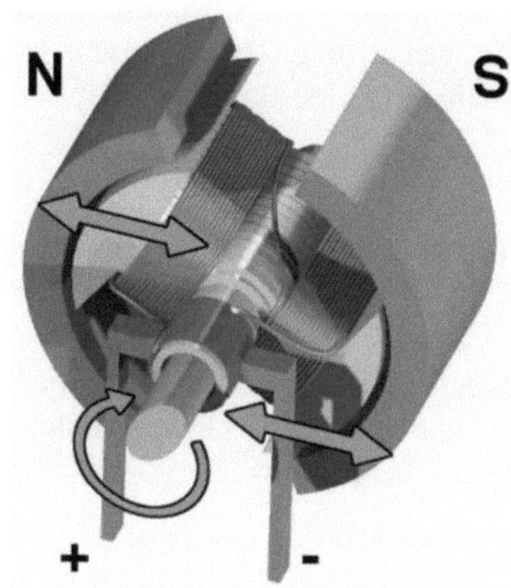

Fig. 10.29 Generator under load. *Source* original from wikicommons under CC3.0 use license [20]

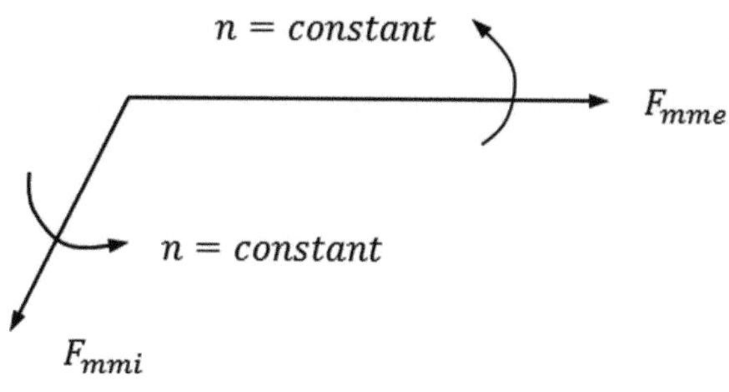

Fig. 10.30 Two rotating vectors. *Source* own elaboration

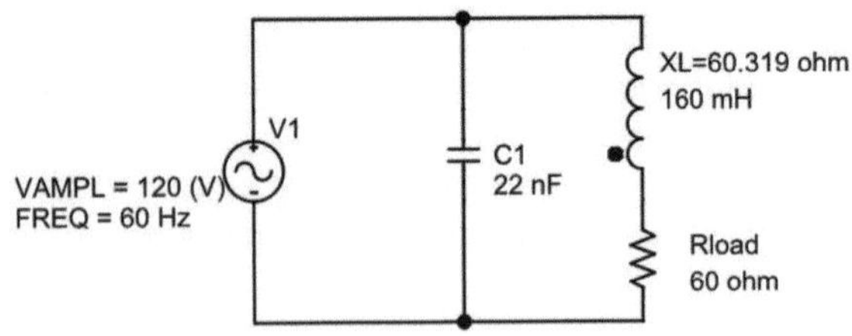

Fig. 10.31 Capacitor in an inductive circuit. *Source* own elaboration.

The reaction in the armature rotates at the same speed and in the same direction as the inductor rotating field that the rotor creates. Then we have two phasors that rotate in solidarity at synchronous speed (Fig. 10.31). Therefore, when a synchronous generator is idle, it is the rotor that creates the only existing rotating field in the machine.

The rotating field induces the emf in the stator windings:

$$E_0 = 4,44 \cdot N \cdot \Phi_0 \cdot f \ per \ phase \tag{10.19}$$

Question What is the emf value that the rotating field induces in the stator windings?

When the generator works under load, the terminal voltage differs from the no-load voltage. This is because the stator windings and the load are both subject to AC current flow. These currents cause the armature reactive field to rotate at the synchronous speed n.

10.7 Power Factor Correction

Question What loads do we normally find in industrial facilities?

Industrial facilities use to employ inductive receivers such as motors, transformers, gas discharge lamps, etc.

Question How do inductive receivers generate the magnetic fields?

These inductive receivers work with considerable reactive power to generate the magnetic fields they need to operate.

Question What effects does the reactive power consumption have?

The consumption of reactive power causes that the power supply company has to provide extra energy that does not make effective work. Because the reactive power is fed back into the grid every ¼ cycle, the electric current in the grid rises. Having a low power factor has the following drawbacks:

For the same useful power (P_u), the current increases as the power factor decreases. As a result, the Joule effect in the wires increases, and then the losses increase as well. Therefore, the efficiency of machines and distribution systems decreases. Recalling that:

$$P = U \cdot I \cdot cos\,\varphi \tag{10.20}$$

$$P = R \cdot I^2 \tag{10.21}$$

A low power factor results in greater voltage drops. This requires more frequent regulation of generators, transformers and lines than would be necessary with a higher power factor. There will be a waste of the generating plant's available energy.

Question Then, what do the companies require from with respect to the power factor?

Because of these disadvantages, electricity supply companies require that their subscribers consume energy with a power factor ≥ 0.55. Distribution companies apply an extra fee for the reactive energy that inductive loads consume in the billing of electrical energy to ensure that users avoid a low power factor.

This extra charge consists of a surcharge of €/kVAr that they add to the billing, whose value depends on the average power factor of the facility. Both the supplying company and the subscribers must work with a power factor as high as possible. However, in some cases it is not always practical to correct the power factor for high values (for example cos φ > 0.95) because the costs are lower than the benefits.

10.7.1 Average Power Factor

Two wattmeters provide readings to determine the power factor. (a wattmeter itself and another one that works as a varmeter). Recalling the power triangle, we have:

$$\cos\varphi = \frac{P}{S} = \frac{P}{\sqrt{P^2 + Q^2}} \tag{10.22}$$

First wattmeter measures the active power. This gives us the active power (P) (wattmeter) and another for measuring the reactive power. This gives us Q (varmeter). Another way to find it is to divide the wattmeter reading ("W") by the total of the voltmeter and ammeter readings. This yields the apparent power ("S").

The capacitors (22 μF) have to be placed in parallel with the network to counteract the excess of reactive (inductive) power consumption (Fig. 10.31). Capacitors absorb the total or partial reactive power of the coils. The voltage source supplies 994.76 mA. Towards the capacitor flow 995.257 mA, and to load it flows 1.41 A.

Then, we plot the voltage and current waveforms (Fig. 10.32). The coils discharge the reactive power to the capacitors in ¼ cycle. During the following quarter cycle, the capacitors discharge the reactive power back into the coils, allowing them to generate their magnetic fields, and the cycle continues. Remarkably, when alternating current supplies energy to the capacitors, different armatures begin to accumulate charge.

By installing a capacitor, the reactive power that the coils produce is lower. Power factor correction lowers the network's reactive power (Q) upstream of our capacitor installation while maintaining the active power (P), which remains constant. As a

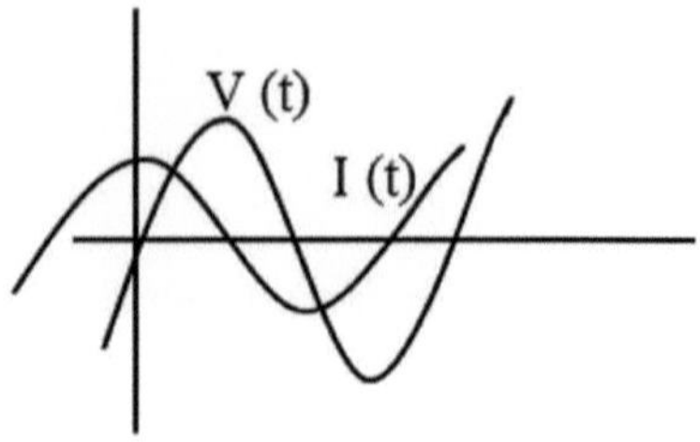

Fig. 10.32 Voltage and current waveform of a purely capacitive circuit. *Source* own elaboration

result, there are several advantages, including a smaller wire section, fewer voltage drops, lower power losses, and more. This also reduces apparent power (S) and current (I).

10.7.2 Capacitor Calculation for Power Factor Correction

Using a specific example, this section examines the behavior of an inductive circuit after adding a capacitor. Suppose we have a 220 V single-phase 60 Hz induction motor that draws 7.6 amperes of current with a power factor of 0.787. Find the size of a capacitor that connects in parallel to turn the power factor to unit. Figure 10.33 the circuit containing the load with the aforementioned power factor. The capacitor corrects the power factor.

Now we assume that in the receiving installation "Z" has the voltage U, and consumes P (kW) of active power, with the power factor (cos φ). Now, we draw the power triangle before adding the capacitor and after adding it (Fig. 10.34).

The difference between the reactive power that the load (Q) consumes and the reactive power that we ultimately want in the circuit yields the value that the capacitor must have.

$$Q_c = Q - Q' \tag{10.23}$$

Be it:

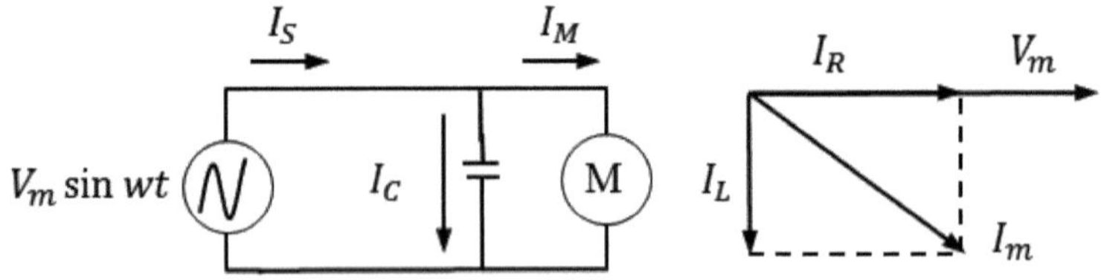

Fig. 10.33 Circuit to correct the power factor and current power factor. *Source* own elaboration

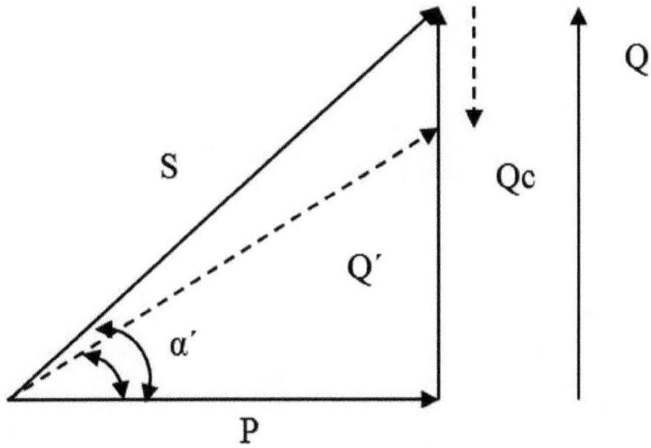

Fig. 10.34 Power triangle diagram. *Source* own elaboration

$$Q = P \cdot tg\varphi \tag{10.24}$$

$$Q' = P \cdot tg\varphi' \tag{10.25}$$

Then, substituting in (Eq. 10.24)

$$Q_c = P \cdot \left(tg\varphi - tg\varphi'\right) \tag{10.26}$$

Remember: $i = C\frac{dv}{dt}$

In the frequency-domain steady state time we can write:

$$I = jwC\ V = \frac{V}{Z_C};\ Z_C = \frac{1}{jwC};$$

Then, the complex power is:

$$S = V \cdot I^* = V\left[\frac{V}{Z_C}\right]^* = \frac{|V|^2}{Z_C^*} = |V|^2 \cdot \left[jwc\right]^* = |V|^2 \cdot \left[-jwc\right] = P + jQ$$

$$P = 0;\quad Q = -w \cdot c \cdot |V|^2$$

Using Eq 10.26 to find the value of the capacitor yields:

$$Q_c = U_c \cdot I_c = U_C^2 \cdot \omega \cdot C \rightarrow C = \frac{Q_C}{U_C^2 \cdot \omega} = \frac{P \cdot \left(tg\varphi - tg\varphi'\right)}{U_C^2 \cdot \omega} \tag{10.27}$$

If:

$$I_c = \frac{U_c}{X_c} \quad (10.28)$$

And:

$$X_c = \frac{1}{\omega \cdot C} \quad (10.29)$$

What is the value of I_c?

Substituting (Eq. 10.28) into (Eq. 10.29).

$$I_c = U_c \cdot \omega \cdot C \quad (10.30)$$

The electric motors operate at 230 V when we wire them in a delta configuration. Operating at this voltage maximizes the efficiency of such a three-phase motor. Recalling the efficiency of a motor:

$$\eta = \frac{P_u}{P_{abs}} \cdot 100 \quad (10.31)$$

P_{abs} is the active electrical power that the motor consumes from the electrical network. P_u is the useful or mechanical power that the motor develops in its shaft, and P_p is the power that the motor losses.

$$P_P = P_{abs} - P_u \quad (10.32)$$

Problem 10.1 Suppose we have a three-phase induction motor with a useful power of 25 hp, an efficiency 75 %, 400 V, and cos fi = 0.6.

Determine the active power as well the current in the wire. Determine the voltage across the terminals of the inductors as well as the current flowing through the coils of the motor if:

1. The motor has star-connected coils.
2. The motor has delta-connected coils.

Question How do we calculate power factor correction?

The element we use to correct power factor is the capacitor, star or delta-connected, which connect to the electrical network that we want to compensate. The most common way is by three-phase automatic capacitor banks. We find the single-phase capacitor.

For the capacitance C we know that:

$$C = \frac{Q_C}{U^2 \cdot \omega} = \frac{P \cdot (tg\varphi - tg\varphi')}{U^2 \cdot \omega} \quad (10.33)$$

Calculation of star-connected capacitors

Suppose we have a facility with a three-phase load in which we intend to place a capacitor bank connected in a star configuration (Fig. 10.35).

We begin with a three-phase scheme, in which we intend to install a star-connected capacitor bank (Fig. 10.36). Recalling (Eq. 10.34) we have

$$Q_C = P \cdot \left(tg\varphi - tg\varphi'\right) \tag{10.34}$$

Because this is a three-phase system, we have to find the capacitance that makes up the star configuration. We proceed by using 1/3 of the total reactive power that the circuit need, based on this premise (Q_C). We can write:

$$\frac{Q_C}{3} = U_F \cdot I_F = U_F^2 \cdot \omega \cdot C_y \rightarrow C_y = \frac{Q_C}{3 \cdot U_{RN}^2 \cdot \omega} = \frac{P \cdot \left(tg\varphi - tg\varphi'\right)}{3 \cdot U_{RN}^2 \cdot \omega} \tag{10.35}$$

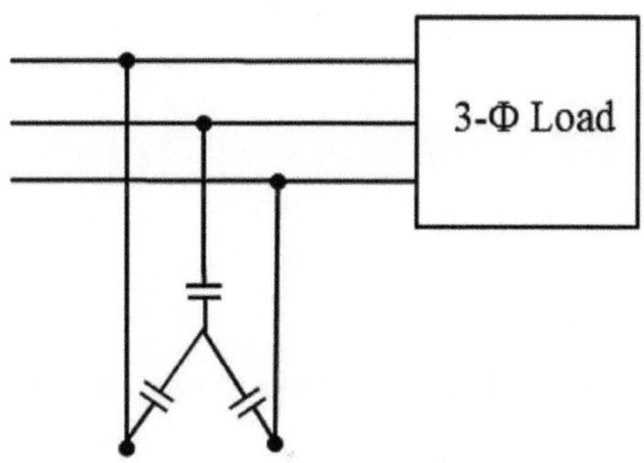

Fig. 10.35 Star-connected capacitors. *Source* own elaboration

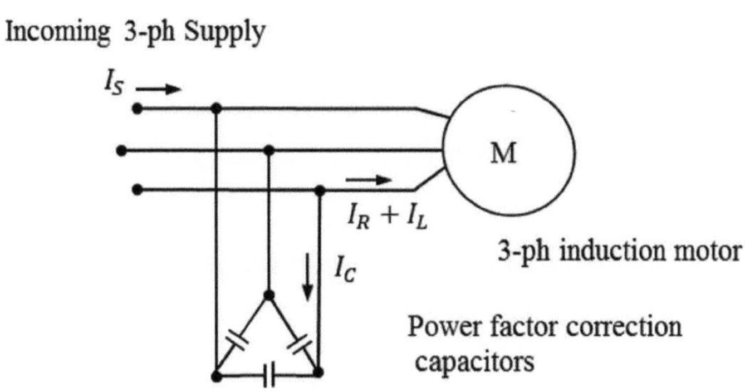

Fig. 10.36 Triangle-connected capacitor. *Source* own elaboration

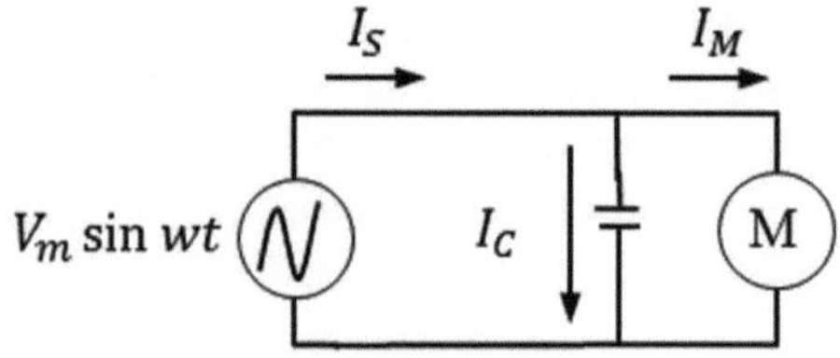

Fig. 10.37 Equivalent circuit of the triangle-connected capacitor. *Source* own elaboration

It is important to note that U_F represents the phase voltage in the star configuration. Precisely, it corresponds with the single-phase voltage of the power supply (U_{RN}). The aim is to place a delta-connected capacitor bank (Fig. 10.36).

Finally, we use the same term as in a single-phase installation.

$$Q_c = P \cdot (tg\varphi - tg\varphi') \tag{10.36}$$

Because this is a three-phase voltage system, our key goal is to compute the capacitor's value in our delta arrangement. This technique makes it easier to find the necessary capacitance in accordance with the specifications of our three-phase system. We will concentrate on one-third of the mandatory reactive power to achieve accurate calculations (Q_C). Then we have:

$$\frac{Q_C}{3} = U_F \cdot I_F = U_F^2 \cdot \omega \cdot C_\Delta \rightarrow C_\Delta = \frac{Q_C}{3 \cdot U_{RS}^2 \cdot \omega} = \frac{P \cdot (tg\varphi - tg\varphi')}{3 \cdot U_{RS}^2 \cdot \omega} \tag{10.37}$$

U_F represents the phase voltage in the delta configuration, equals the line voltage of the power supply (U_{RS}).

Problem 10.2 Suppose we have a motor with 5 kW of power and a power factor of 0.75 that connects to a 230 V source, 50 Hz single-phase network (Fig. 10.37). Determine the characteristics of the capacitor bank that we require to enhance the power factor to 0.95.

Find:

(a) The network current once we place the capacitor bank.
(b) The current through the capacitors.

10.7.3 Type of Reactive Power Compensation

Capacitors are parallel-connected to the load that requires compensation in order to achieve reactive power compensation. In an electrical facility with several inductive loads, there are two main techniques to perform the compensation. First, the reactive power compensation using delta-connected capacitors (Fig. 10.38).

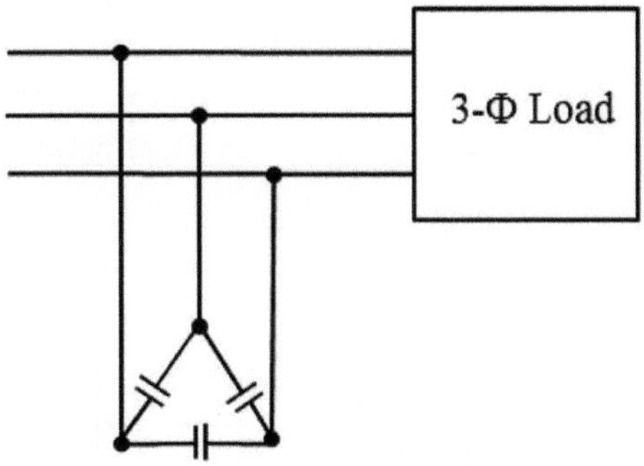

Fig. 10.38 Delta-connected capacitors. *Source* own elaboration

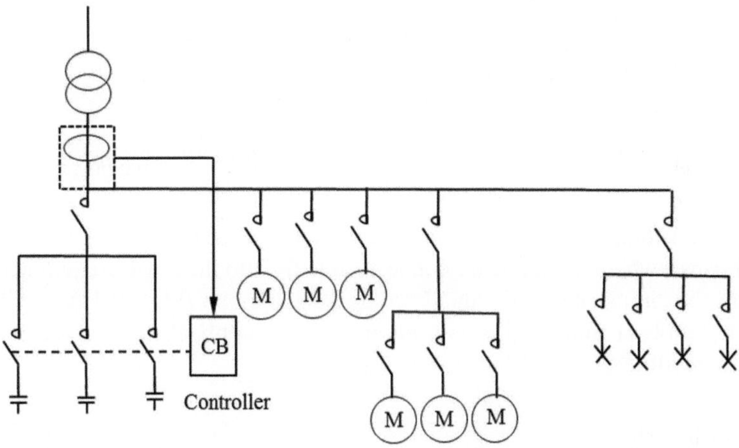

Fig. 10.39 Capacitors connected to compensate reactive energy. *Source* own elaboration

10.7.3.1 Single Reactive Power Compensation

In order to perform single compensation, we connect the capacitors in parallel with the inductive load to compensate reactive power. It is a circuit that compensates the inductive energy that motors, or others components provoke (Fig. 10.39).

10.7.3.2 Central Compensation with Capacitors

The reactive power regulator is the device that measures the compensation factor of the system continuously and is capable of connecting and disconnecting the capacitors (Fig. 10.40).

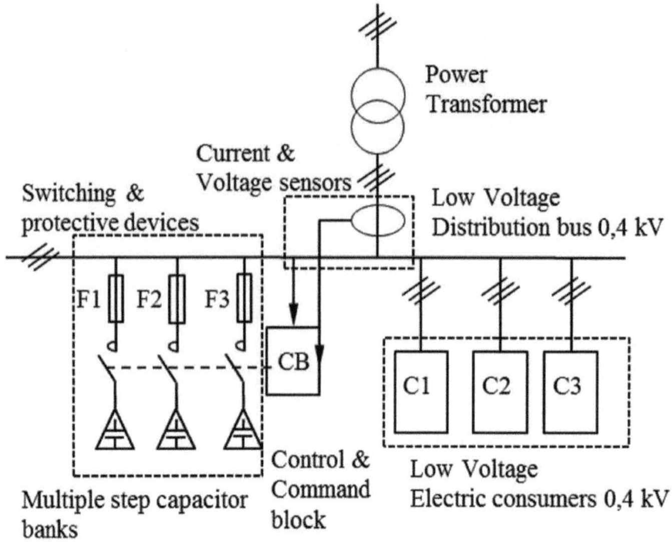

Fig. 10.40 Automatic system for power factor correction. *Source* own elaboration

Individual compensation is a costly measure that we use for discharge lamps (high factor lamps), where the capacitor connects directly into the starting equipment. We can draw the circuit of an individually compensated motor (Fig. 10.41).

Central/global compensation with automatic capacitor banks works better in large industrial facilities.

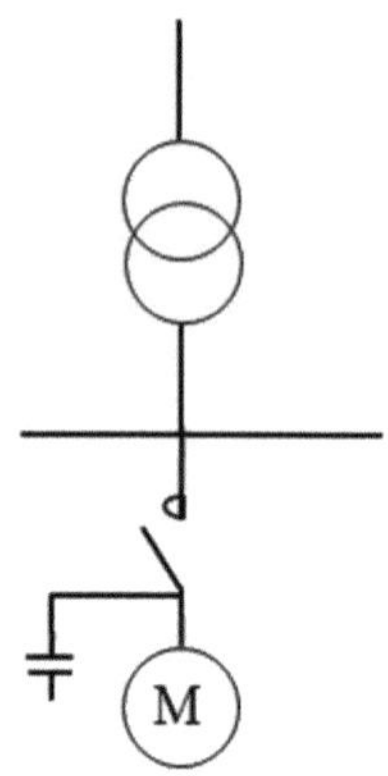

Fig. 10.41 Individual compensation of a motor. *Source* own elaboration

10.7.4 Multi-receiver Systems

For both single-phase and three-phase installations, where multiple loads connect to the same installation, it is essential to find the total installed power, power factor, and total current. In these, it is true that:

$$P_T = P_1 + P_2 + P_3 + .. \tag{10.38}$$

$$Q_T = Q_1 + Q_2^2 + Q_3 + .. \tag{10.39}$$

$$S_T = \sqrt{P_T^2 + Q_T^2} \tag{10.40}$$

The procedure is the same for single-phase and three-phase loads. In this case, we determine the total active and reactive power of all loads. Upon obtaining both the active and reactive power, total apparent power, power factor, and current will all be found out.

Question What would occur with capacitive loads?

We deduce the reactive power they generate from that of their inductive counterparts. This applies to single-phase systems.

Recalling that:

$$\mathrm{S_T} = \mathrm{U} \cdot \mathrm{I_T} \rightarrow \mathrm{I_T} = \frac{\mathrm{S_T}}{\mathrm{U}} \tag{10.41}$$

$$\cos \varphi_T = \frac{P_T}{S_T} \tag{10.42}$$

For a three-phase system we know that:

$$\mathrm{S_T} = \sqrt{3} \cdot \mathrm{U_C} \cdot \mathrm{I_L} \rightarrow \mathrm{I_L} = \frac{\mathrm{S_T}}{\sqrt{3} \cdot \mathrm{U_C}} \tag{10.43}$$

$$\cos \varphi_T = \frac{P_T}{S_T} \tag{10.44}$$

It is worth noting that, in a three-phase system, each phase of the capacitor bank will wield one third of the total power. It should be kept in mind that:

$$C = \frac{P \cdot (tg\varphi - tg\varphi')}{3 \cdot \omega \cdot U_C^2} \tag{10.45}$$

In a star-arrangement:

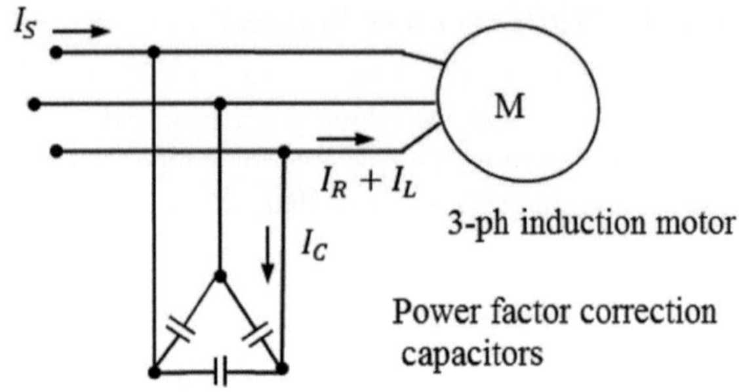

Fig. 10.42 Problem 10.3. *Source* own elaboration

$$Q_C = U_S \cdot I_{fc} \tag{10.46}$$

$$I_{fc} = \frac{U_S}{X_C} \tag{10.47}$$

In a triangle-arrangement:

$$Q_C = U_C \cdot I_{fc} \tag{10.48}$$

$$I_{fc} = \frac{U_C}{X_C} \tag{10.49}$$

Problem 10.3 Consider the three-phase induction motor (Fig. 10.42). The motor has the following parameters: useful power: 20 HP; Efficiency ($\eta = 75\ \%$); Voltage (400 V), and $\cos\varphi = 0.7$.

Determine the dimension of the wye capacitor bank to be parallel-connected to the motor to enhance the power factor of 0.8. Calculate the delta capacitor bank that must be parallel-connected to the induction motor to enhance a power factor of 0.8.

10.8 Summary

This chapter provides a detailed examination of the synchronous machine, a key component in power generation and large industrial motors. It begins by describing its constituent elements, including the stator, rotor, windings, excitation system, and cooling system.

The core of the chapter explains the machine's operation, starting with the principles of an elementary synchronous generator and progressing to a real synchronous generator under both no-load and loaded conditions. A significant focus is placed on the machine's role in power factor correction, detailing how to calculate the required capacitors for compensation and discussing different compensation strategies, such as single and central reactive power compensation for multi-receiver systems.

References

1. Biezl. Schenkelpolmaschine (2009). https://commons.wikimedia.org/wiki/File:Au%C3%9Fenpolmaschine.svg. Last accessed 2 de enero de 2024
2. A. Ситенький, English: A generator at the end of a steam turbine at the Balakovo Nuclear Power Plant (2008). https://commons.wikimedia.org/wiki/File:BalakovoNPP_tb.jpg. Last accessed 7 Dec 2023
3. D. Huebler, Electrical Fault in stator winding of no. 2 drive motor of 10 × 10 foot wind tunnel secondary drive system (1977). https://commons.wikimedia.org/wiki/File:ELECTRICAL_FAULT_IN_STATOR_WINDING_OF_NO._2_DRIVE_MOTOR_OF_10X10_FOOT_WIND_TUNNEL_SECONDARY_DRIVE_SYSTEM_-_NARA_-_17468957.jpg. Last accessed 2 de enero de 2024
4. Wikipedia. Stator machine triphasée.jpg. https://commons.wikimedia.org/wiki/File:Stator_machine_triphas%C3%A9e.jpg. Last accessed 22 Dec 2023
5. Biezl. Torquemotor als Innenläufer in Wellenwicklung (2017). https://commons.wikimedia.org/wiki/File:Torque_motor.svg. Last accessed 7 Dec 2023
6. Wikipedia. Alternator. https://commons.wikimedia.org/wiki/File:Alternador.svg. Last accessed 7 Dec 2023
7. U. Jm. modification of original svg by User:SiriusA, Three Phase alternating current waveform. This figure illustrates one voltage cycle of a three-phase system, labeled 0 to 360° (2 π radians) along the time axis. The plotted line represents the variation of instantaneous voltage (or current) with respect to time. This cycle will repeat 50 or 60 times per second, depending on the power system frequency. The colors of the lines represent the American color code for 120v three-phase. That is black=VL1 red=VL2 blue=VL3. 2008. https://commons.wikimedia.org/wiki/File:3_phase_AC_waveform.svg. Last accessed 2 de enero de 2024
8. C.E. Magnussen, English: Bars and rings of the damper (amortisseur) winding of an AC generator (General Electric, early 20th century). Notice the gaps in the cage along the quadrature axes (1916). https://commons.wikimedia.org/wiki/File:Damper_winding.png. Last accessed 6 Dec 2023
9. K./D. Gubler, Deutsch: Gleichstromversorgung der Erregerwicklung eines Synchrongenerators im Kraftwerk Amsteg (2007). https://commons.wikimedia.org/wiki/File:Amsteg-laeuferversorgung.jpg. Last accessed 6 Dec 2023
10. Kabelleger, Erreger (links vorne) und Generator (hinten)/Exciter (front left) and generator (back) (2007). Accessed on 2 de enero de 2024. https://commons.wikimedia.org/wiki/File:Amsteg-stillgelegte-generatorgruppe-erreger.jpg
11. Haade. 2 Pole DC motor PM (2007). https://commons.wikimedia.org/wiki/File:Motor_DC_2Pole.png. Last accessed 6 Dec 2023
12. Svjo, English: Rotating 3-phase magnetic field. https://commons.wikimedia.org/wiki/File:Rotating-3-phase-magnetic-field.svg. Last accessed 6 Dec 2023
13. Biezl. Schenkelpolmaschine. https://commons.wikimedia.org/wiki/File:Innenpolmaschine.svg. Last accessed 6 Dec 2023
14. Biezl. Vollpolmaschine (2009). https://commons.wikimedia.org/wiki/File:Vollpolmaschine.svg. Last accessed 6 Dec 2023
15. Wikipedia. Rotating magnetic field. https://commons.wikimedia.org/wiki/File:Rotating_magnetic_field.png. Last accessed 6 Dec 2023
16. Wikipedia. Diagram of single-phase generator with four poles.svg - Wikipedia». https://commons.wikimedia.org/wiki/File:Diagram_of_single-phase_generator_with_four_poles.svg. Last accessed 6 Dec 2023

17. Wikipedia. Diagram of single-phase generator with two poles.svg. https://commons.wikimedia.org/wiki/File:Diagram_of_single-phase_generator_with_two_poles.svg. Last accessed 6 Dec 2023
18. Wikipedia. Polyphase system, Accessed on 6 March 2026. [Online]. Available at: By Nehemiah Hawkins-File:Hawkins Electrical Guide-3phase Elementary 6wire.jpg, Public Domain, https://commons.wikimedia.org/w/index.php?curid=182146835
19. Zureks. Stator of an electric motor with windings (2006). https://commons.wikimedia.org/wiki/File:Uzwojenie.jpg. Last accessed 6 Dec 2023
20. Wikipedia. Electric motor cycle 2.png. https://commons.wikimedia.org/wiki/File:Electric_motor_cycle_2.png. Last accessed 6 De 2023

MIX
Papier aus verantwortungsvollen Quellen
Paper from responsible sources
FSC® C105338

If you have any concerns about our products,
you can contact us on
ProductSafety@springernature.com

In case Publisher is established outside the EU,
the EU authorized representative is:
Springer Nature Customer Service Center GmbH
Europaplatz 3, 69115 Heidelberg, Germany

Printed by Libri Plureos GmbH
in Hamburg, Germany